Sustainable Agriculture

NIPA® GENX ELECTRONIC RESOURCES & SOLUTIONS P. LTD.
New Delhi-110 034

Sustainable Agriculture

— A Vision for Future —

B.K. DESAI
Associate Professor, Department of Agronomy
B.T. PUJARI
Professor and Head, Department of Agronomy and
Associate Director of Research

College of Agriculture, Raichur - 584 101 Karnataka

NIPA® GENX ELECTRONIC RESOURCES & SOLUTIONS P. LTD.
New Delhi-110 034

NIPA® GENX ELECTRONIC RESOURCES & SOLUTIONS P. LTD.

101,103, Vikas Surya Plaza, CU Block
L.S.C. Market, Pitam Pura, New Delhi-110 034
Ph : +91 11 27341616, 27341717, 27341718
E-mail: newindiapublishingagency@gmail.com
web: www.nipabooks.com

For customer assistance, please contact
Phone: + 91-11-27 34 17 17
Fax: + 91-11- 27 34 16 16
E-Mail: feedbacks@nipabooks.com

ISBN: 978-93-58870-16-9

Composed and Designed by NIPA®.

FOREWORD

The pressing need to feed our large population has led to over exploitation of land resulting in its enormous degradation, which is a major concern in present day land-use management. The objective of food security can be met by promoting sustainable agricultural production systems. Increasing agricultural production should not be at the cost of degrading soil and depleting water bodies.

The agricultural research faculty is forced with the challenge of developing sustainable systems in order to maintain the goal of self sufficiency in long run.

In this context, I feel, it is the right time to bring out a publication on "Sustainable Agriculture - A vision for future". It covers all aspects of sustainable agriculture which are judiciously incorporated in various chapters. The authors have compiled most of the useful information from the available sources which has made the book a highly valuable asset for the readers.

I heartily congratulate the authors for their sincere efforts in bringing out the publication and I hope that the book will be widely referred by students, researchers and farmers as well.

Sd/-

20-4-2007

Dr. J. H. KULKARNI

Vice-Chancellor

University of Agricultural Sciences, Dharwad

PREFACE

There is an old proverb "Words fly away; writing remains," i.e. what we put down on paper may not fade away soon; these words may last longer and go much farther, with this intention, we have made an attempt to bringout this book in its present form. There is lot of confusion in alternative agriculture systems being promoted in India and elsewhere. Further, the available sources of literature on sustainable agriculture are so diverse that it becomes difficult to locate the required information in one place. Several books on organic farming written by eminent authors are available. However, the approach made in this book is entirely different, here sustainable agriculture is viewed as a complete umbrella term and other alternative systems are brought under it. Another unique feature in this book is the blending of information of Indian and foreign workers, keeping in view the requirement. Moreover, a compilation of available information on sustainable Agriculture has been a felt need of students, teachers & researchers.

Hence, we have made an attempt to present available information in an easily understandable manner hoping that, bringing information together in one volume will contribute to the advancement of knowledge.

It is to mention here that, the authors don't claim credit for original contribution on some of the topics, rather it should be viewed as collection of various information from different books, journals, websites, articles etc., in a presentable manner under one roof. In-most of the cases we have quoted the authorship even then, if there are some omissions, may be in the text topics or in the references it is not intentional. Further, it is very difficult to name and acknowledge each and every references used. We sincerely thank the authors for whatever help we have received In compiling this book.

We hope the contents of the book will be of immense value to students (both UG & PG), teachers, research workers and extension personnel. We also hope that the readers will appreciate the fluency regarding the arrangement of topics and the subject matter. Any constructive suggestions and criticisms are most welcome.

We place on record our profound indebtedness to our honourable Vice-Chancellor Dr. J. H. Kulkarni for providing foreword. Further, we thank our publishers for bringing out this book in an effective way, we also thankfully acknowledge the typing help received by Sri. Ramakrishna Shirol and Sri Shafee Sullad.

Raichur
25-06-2007

B.K. Desai
B.T. Pujari

CONTENT

1 INTRODUCTION

Preserving the productivity of agricultural land over a long term requires sustainable food production. Sustainability is achieved through alternative agricultural practices informed by in-depth knowledge of the ecological processes occurring in farm fields and the larger landscape of which they are a part. From this foundation, we can move towards the social and economic changes that promote sustainability of all sectors of the food system. Sustainability means different things to different people, but there is a general agreement that it has an ecological basis.

In recent years, all eyes seem to be focused on visions of the future. High-tech, bio-tech, and info-tech are the key words most frequently used among those who are concerned with future agriculture development. Many also forecast increasing vertical integration of production, processing distribution functions-including activities from molecular level to consumer preference i.e. from seed to table, thus increasing dependence on biotechnology and information technology at all levels in world agriculture.

Conventional farming systems vary from farm to farm and from country to country. However, they share many common characteristics *viz.,*rapid technological innovation;

large capital investments in order to apply production and management technology; large-scale farms; single crops/ row crops grown continuously over many seasons; uniform high-yielding hybrid crops; extensive use of pesticides, fertilizers, and external energy inputs; high labour efficiency; and dependency on agribusiness. In the case of livestock, most production comes from confined, concentrated systems.

Agricultural development in independent India can be divided into 3 distinct phases. The phase I started during 1950, wherein local cultivars, indigenous technology, green manuring being the inputs and the expected output was quite low. Alarming rise in the population rate forced the scientists to think of alternatives to the local cultivars with higher response to fertilizers & jump in yield levels. This marked the beginning of phase-II (1965), wherein high yielding varieties, fertilizers, mechanistion & irrigation were used as the inputs and production increase was lucrative. And this phase was popularly called as 'Green Revolution' phase in the History of Indian Agriculture. But after 20-25 years the impact of high input technology was reflected in the degradation of agro-ecosystem & once again forced the scientists to think of a viable option to keep intact the resource base & to harvest more/sustain yield levels to feed the burgeoning population. This being the third phase is popularly called as phase of sustainable agriculture organic farming phase started from 1980's onwards. This gave an impetus to environmentalists, to think seriously regarding protection of the ecosystem.

The main cause for degradation of the ecosystem was thought to be the green revolution period, during which more emphasis was given to the use of high yielding varieties. Agriculture has been & will always remain the most important sector of a national and global economy which provides the ultimate in essential food & fibre for the world's population. No Industrial substitutes have yet been found to replace food requirement. Thus the long term survival of mankind, will depend on the sustainability of the global agricultural economy. A key factor in farmer's acceptance.

Sustainability should be visualised as an ultimate goal and it should be activated in all walks of life through various production practices.

India's Foodgrain Needs

In 1951, the population was only 361 million people and by 2005 the population reached 1000 million nearly trebling our population in the second half of twentieth century. A well-planned and concerted effort by agricultural scientists, extension workers, farmers and the government price support policies have made 'green revolution' realizable by increasing grain production at a rate faster than the increase in population. This has kept us away from hunger despite some very bad drought years. Hungry people can hardly think of a quality environment. Three major inputs that have made it possible are: seeds of high yielding varieties of crops, fertilizer and increased irrigation facilities. But then it is not all over and although there are good signs of decline in the rate of population increase from 2.14 to 1.70%, we are still likely to add another 420.5 million people in the next 20 years by 2020 AD, that is about 21 million people each year. This will certainly increase our food demand.

The total foodgrain demand by 2020 is estimated at 294 MT (122 MT rice, 103 MT wheat, 41 MT coarse grains and 28 MT pulses). Thus by 2020 we need to produce about 100 MT of additional food grain per year from the same or even less area (some more area will go to meet the increasing needs for roads, rails, buildings, etc.). We have no choice but to increase the fertilizer application. During 1980–90 there has been 3–4% decrease in fertilizer nitrogen consumption in Europe and USA, while in Asia it has increased by 74.4%. In India the current fertilizer consumption is much below the mark. Only 19 out of 437 districts in India consumed more than 200 kg $N + P_2O_5 + K_2O$ ha^{-1} in 1996-97, while 176 districts consuming only 50 kg ha^{-1} or less.

Threats to Sustainability

- Population
- Limited resources - natural or artificially expanded
- Wastage

With increasing global human population, there is growing awareness of the need to increase food production while protecting biodiversity and the natural environment. Humans have the opportunity and responsibility to evaluate food systems in new ways, to recognize the need to balance the system with available resources, and to accept a moral obligation to manage outputs from the system in an equitable manner. When people are viewed as an integral part of the ecosystem, subject to all the natural laws and consequences of system success, there is a compelling reason to make agroecosystems as sustainable as possible for the long term. Beyond our current disruptive power in the ecosystem, we are capable of designing systems that close nutrient cycles, depend more on renewable energy, reduce inefficiencies in production, and promote environmental health. We can meet our goals by using some system design principles and properties that resemble those of natural ecosystems.

In this context, agroecology has appeal because it helps us focus on structure and processes at each relevant systems level. Careful study can lead to better analyses of the sustainability or potential negative long-term environmental impacts of current agricultural practices and systems. There are growing concerns by the public, often expressed through regulations, that lead to requirements for agricultural practices that will reduce the loss of soil and nutrients from fields and minimize the entry of pesticides and their residues into surface and groundwater. Learning more about the cycling processes and designs of natural system can help us improve managed agroecosystems. Up to the present, the focus in agricultural science has been primarily on components of the production process, and maximizing net returns of single products per unit of land or labor. All other resource use and environmental effects have been considered "externalities," and have been excluded from system design. The production focus is also reflected in design of most current agroecology courses.

The agricultural system is an open system, interacting with nature and with society, and the development of a sustainable food system will require more attention to the efficiency of the entire process of converting natural resources to what reaches consumers' tables. This includes analysis of food production, processing, marketing, and consumption. When agroecology is defined as the ecology of food systems, we are obligated to look at more than the efficiencies of resource use in production, the short-term environmental impacts of practices, and annual enterprise economics. Most of the energy (perhaps >75%) in the food system involves steps after the field production process (Johansson *et al.*, 2000). We need to consider the energy used and waste generated at each step in the food chain, the potentials for cycling materials back into primary production, and the emergent properties of a complex system that is basic to human survival. We need to employ such tools as materials life cycle analysis (Audsley *et al.*, 1997), energy analysis (Odum, 1996), environmental footprint calculations (Wackernagel *et al.*, 1999), and alternative economic and other valuation schemes (for review see Doherty and Rydberg, 2002). Beyond looking at only the energy and materials flows, we must consider other driving forces in the system such as economics at the farm, national, and global levels, the environmental consequences of systems on all plant and animal species, and the social and health impacts of systems on people. An interdisciplinary, integrated approach is essential to adequately address the complexities of interactions in the total food system (Allen *et al.,* 1991).

The recent intensification of agriculture, and the prospects of future intensification, will have major detrimental impacts on the non agricultural terrestrial and aquatic ecosystems of the world. The doubling of agricultural food production during the past 35 years was associated with a 6.87-fold increase in nitrogen fertilization, a 3.48-fold increase in phosphorus fertilization, a 1.68-fold increase in the amount of irrigated cropland, and a 1.1-fold increase in land in cultivation. Based on a simple linear extension of past trends, the anticipated next doubling of global food production would be associated with approximately 3-fold increases in nitrogen and phosphorus fertilization rates, a doubling of the irrigated land area, and an 18% increase in cropland. These projected changes would have dramatic impacts on the diversity, composition, and functioning of the remaining natural ecosystems of the world, and on their ability to provide society with a variety of essential ecosystem services. The largest impacts would be on freshwater and marine ecosystems, which would be greatly eutrophied by high rates of nitrogen and phosphorus release from agricultural fields. Aquatic nutrient eutrophication can lead to loss of biodiversity, outbreaks of nuisance species, shifts in the structure of food chains, and impairment of fisheries. Because of aerial redistribution of various forms of nitrogen, agricultural intensification also would eutrophy many natural terrestrial ecosystems and contribute to atmospheric accumulation of greenhouse gases. These detrimental environmental impacts of agriculture can be minimized only if there is much more efficient use and recycling of nitrogen and phosphorus in agroecosystems.

Almost 70 per cent of all available freshwater is used for agriculture. Over pumping of groundwater by the world's farmers exceeds natural replenishment by at least 160 billion cubic metres a year. It takes an enourmous amount of water to produce crops: one to three cubic metres to yield just one kilo of rice, and 1,000 tons of water to produce just one ton of grain.

Land in agricultural use has increased by 12 per cent since the 1960s to about 1.5 billion hectares. Current global water withdrawals for irrigation are estimated at about 2,000 to 2,555 km^3 per year. Pasture and crops take up 37 per cent of the Earth's land area.

Poor drainage and irrigation practices have led to saline build-up in about 30 million hectares of the world's 240 million hectares of irrigated land, according to the UN Food and Agriculture Organization (FAO). A combination of salinization and waterlogging affects another 80 million hectares.

Need

Needs of sustainable agriculture

"A farmer should live as though he were going to die tomorrow;

but he should farm as though he were going to live forever"

East Anglian proverb, in George Ewart Evans, 1966

Conditions of sustainable agriculture with the continued degradation of land and natural resources (Water, Soil and genetic resources) increasing attention is directed to sustainable agriculture. There is now an urgent demand for creative and innovative conservation and production practices that would provide farmers with economically viable and environmentally sound alternatives in their agricultural production systems.

Sustainable agriculture encompasses the elements of productivity, profitability, conservation, health safety and the environment.

Food and nutritional security seems to be the serious global conservation on a global scale, agriculture has been very successful in meeting a growing demand for food. The rate of increase in food production has generally exceeded the rate of population growth and has disproved the productions of Paul ehrlich "Some time between 1970 and 1985 the world would undergo vast famines wherein hundreds of millions of people are going to starve to death. The post independence India with population increasing at alarming rate was put to threat of insufficient food production. But the vision and collective efforts of planners, scientists and farmers proved these predictions wrong leading to self sufficiency.

Despite the phenomenal growth in food production since the mid sixties, some 800 million people in the world still suffer from chronic hunger. This situation has arised because of the uneven spread of natural and economic wealth and also because of modern production systems in India and around the world that has grossly overlooked the issues of environmental sustainability and social equity.

Thus there is global awareness with respect to health hazards and environmental concern is gaining momentum and scientists, planners and farmers are in the quest for searching a viable alternative.

The United States Department of Agriculture's report and recommendation (USDA 1980) has listed the concerns often expressed to the USDA study team *viz.,*

- Increased cost and dependence on external inputs of chemicals and energy.

- Continued decline in soil productivity from excessive soil erosion and nutrient run off losses.
- Contamination of surface and ground water from fertilizers and pesticides
- Hazards to animal health and to food quality and safety from agricultural chemicals.
- Fading out of the family farms and localized marketing systems.

These concerns, have paved way for an alternative system of farming with emphasis on productivity, stability, profitability, environmentally served thus tending towards sustainability. But there is a big difference between advanced countries and developing countries when addressing the problems in the background of sustainable agriculture.

The necessity is because of environmental pollution caused by chemical fertilizers and agricultural chemicals, the safety of food are the major reasons.

The background in developing countries is entirely different wherein the major reasons are, the reduction of productivity and desertification caused by over exploitation of resources, the accumulation of salts in the soil caused by irrigated agriculture, the destruction of soil and the decrease in sustainability all these have led to the need of sustainable agriculture.

The safety of food is being proposed from the viewpoint of maintaining health between producers and consumers. This problem, including global environmental problems, are the major issues concerned.

Thus, sustainable agriculture is the need of the hour because of the urgency to develop farming techniques, which are sustainable from environments, production and socio-economic point of view.

Also to avoid deleterious effects of synthetic chemical fertilizer and pesticides, sustainable agriculture is needed to provide ecologically safe methods of farming.

Scope

Agriculture faces a new challenge in the coming century, to feed more people on less land, without degrading the natural resource base. The very definition of sustainable agriculture points to a major problem with conventional agriculture, in that it is inherently unsustainable in the long run. There are many reasons why conventional agriculture are unsustainable, but most of them are due to the focus on short term profit that the huge majority of firms in a capitalist world system have. This focus on short term profits has been different facets, and has been exponentially increased with the development of multinational agricultural corporations and the decline of the family farmer.

Since the 1920s when chemical fertilizers were first used commercially on a large scale, there have been claims that agricultural chemicals produce unhealthy and less nutritious food crops. By the 1940s, the organic farming movement had begun, in part due to this belief that food grown using more traditional, chemical-free methods was more healthy. Foods grown by these methods are to be known as "organic." Today, this notion has continued in the alternate health arena, and some alternative treatments, such as the Gerson cancer therapy, rely on food grown organically. But the question remains: whether organically grown food is more nutritious? There are several reasons why there has not been a solid answer to this question. The first is that the difference in terms of health effects is not large enough to readily apparent. In other words, if people stayed well on an organic diet but got violently ill as a result of consuming food grown with chemical fertilizers and pesticides, then the difference would be perfectly obvious; however, this is not the case, and a more subtle difference, such as an 8% increase in the incidence of allergies, for example, is much more difficult to detect and easier to overlook.

A second reason is that it is difficult to conduct and interpret agricultural research investigating nutrient content. Factors such as sunlight, temperature and rainfall, which influence the nutrient content of a crop can occur during storage and shipping. For these reasons, it is difficult both to plan effective studies and to make sense of the results. Furthermore, these considerations often make it difficult to compare the results of different studies. Finally, many of the studies that have been done are relatively old and not performed according to modern standards. In particular, the older studies do not include a rigorous statistical analysis. This factor alone can make these studies difficult to evaluate. As a result, they have been dismissed by some as valueless.

In more than 300 comparisons performed in these studies, organic crops had a higher nutrient content about 40% time. Overall, organic crops had an equal or higher nutrient content about 85%. These results suggest that, on an average, organic crops have a higher nutrient content.

While the overall outlook is favourable for organic crops, there is too little data for most individual nutrients to say anything at all. But for three individual nutrients - vitamin C, nitrates and protein quality - there is enough evidence to suggest that organic crops are superior to conventional ones. Compared to crops grown with chemical fertilizers and pesticides, organically grown crops generally have a higher vitamin C content, a lower content of carcinogenic nitrates and better protein quality. Further work is needed on other nutrients before any definitive conclusion can be drawn.

There are 14 such animal studies that have been performed over the last 70 years. In ten of these, the organically fed animals fared better, in one, the animals fed organic feed came in second among several chemically fertilized feeds; and 3 studies showed no difference, possibly due to weaknesses in the study designs.

Sustainable agriculture is more profitable in terms of money and soil conservation in the long run. Without doubt, it can meet the requirements of the country. Prakruti [a non-governmental organisation Kisan Mehta is associated with] tried to study this issue in the earlier years but very few farmers follow the whole set of practices required in sustainable agriculture. Sustainable agriculture means not only the withdrawal of three things - synthetic chemicals, hybrid-genetically modified seeds and heavy agricultural implements; it is an elaborate system that tries to simulate the conditions found in nature. Multiculture, intercropping, use of farmyard manure and remnants, mulching and application of integrated pest management. If this is followed then there is no reason why agriculture cannot be economically viable in addition to being environmentally sustainable.

The abrupt withdrawal of all this would reduce the yields initially but within three years normal yields can be got. After this yields will keep increasing. Multiculture and intercropping as against monoculture followed in commercial farming, also provide a cushion against money loss. Viability is misunderstood. Also, farmers do not consider the invisible savings in money and labour as a benefit.

There is no question of sustainable agriculture not being viable or not meeting the needs of the country for food and fibre. Following U.S. sanctions, Cuba was left with no option other than to have recourse to sustainable practices. Its agriculture is now fully organic.

A recently published review of 41 scientific studies from countries around the world comparing the nutrition of organic an conventionally grown foods found significantly higher nutrients in organic crops.

"Nutritional Quality of Organic Versus Conventional Fruits, Vegetables, and Grains", published in The Journal of Alternative and Complimentary Medicine, found organic crops, on an average, contained 29.3% more magnesium, 27% more vitamin C, 21% more iron, 13.6% more phosphorus, 26% more calcium, 11% more copper, 42% more manganese, 9% more potassium and 15% lower nitrates. In crops such as spinach, lettuce, cabbage and potatoes, organically grown crops showed even higher nutritional superiority.

Organic foods are the fastest growing and most profitable segment of American agriculture, according to government statistics. A February 1997 poll by the genetic engineering corporation Novartis found that 54% of US consumers would prefer to see organic agriculture becomes the predominant form of food production - as opposed to conventional, chemical-intensive farming or agricultural biotechnology. A June 2000 survey carried out by the National Center for Public Policy, a conservative think tank, indicated that 68-69% of the American public believe that the organic label on food products means that they are safer and better for the environment. This is the main reasons why 10 million organic consumers will buy eight billion dollars worth of organic food this year in the US. In Europe trends indicate that 30% of all farming may be organic by the year 2010. More and more health and environmentally conscious Americans are turning to organic food and for good reason:

- Concern over toxic pesticide residues. A March 1999 study by Consumer Reports found that organic foods had little or no pesticide residues compared to conventional produce. A 1999 study by the Environmental Working Group found that millions of US children eating non-organic fruits and vegetables were ingesting dangerous amounts of a variety of pesticide neurotoxins and carcinogens.

- Concern over food poisoning, deadly E-coli 0158:H7, campylobacter, salmonella, listeria, and other food borne diseases. The Centers for Disease control admit that there are at least 76 million cases of food poisoning every year in the US. While there are no documented cases of organic meat or poultry setting off food poisoning epidemics, filthy slaughterhouses, contaminated feed, and diseased animals are commonplace in industrial agriculture. According to government statistics, most non-organic beef cattle are contaminated with E-coli 0157:H7:over 90% of chickens are tainted with campylobacter, and 30% of poultry are infected with salmonella.
- Concern over food irradiation, use of toxic sewage sludge spread on farmland, and genetic engineering. Organic certification prohibits irradiation, sewage sludge, and genetic engineering. A 1997 poll by CBS found 77% of Americans opposed to food irradiation, while a recent survey by the Angus Reid polling group found the majority of US consumers opposed to genetically engineered foods. Consumers are especially incensed that industry and the FDA refuse to require labelling of generally engineered food. Numerous polls over the past 15 years have found that 80-95% of Americans want labels on gene-altered foods, mainly so that they can avoid having them.
- Concern over the environment. Studies indicate that the industrialization and globalization of agriculture are a leading contributor to greenhouse gases and climate destabilization. Other research shows an increasing percentage of municipal water supplies are contaminated by pesticide residues, chemical fertilizers, and sewage runoff from factory farms and feedlots.

Organic food producers and processors have already been successful in getting information to the public about the certified labels that identify their products in the market, as well as making direct connections for sale from the farm. Just as commodity groups have promoted their specific crop or animal product, organic growers and processors have pursued public education campaigns using grower fees and contributions. One result is that growers benefit from the premium price that organic products will bring in the marketplace, thus increasing incomes by advancing their labeled products. Organic food sales have increased by about 20% per year for more than a decade in U.S. and Europe (Lampkin, 1999).

Emphasis on consumer education should include explanations of how ecological approaches used by sustainable farmers can produce food without pesticides and with alternatives to chemical fertilizers, while protecting the ecosystem and producing healthy, safe food for consumers. The farmer can be portrayed as an environmental steward who is also concerned about consumer health, an ally of the consumer who has made changes in farming practices to achieve these multiple goals. Customers can be affirmed when they make the choices to purchase food produced in this manner. Further education about how food is produced, where, by whom, and under what conditions will reinforce their decisions. Surveys in Norway have shown that consumers who regularly purchase ecological food are also concerned about overall pesticide and fertilizer use, about animal welfare, and about the conditions of farmers and families where food is produced (Torjusen *et al.*, 2001). When agroecology studies the entire food system, these issues of consumer opinion and choice are part of the equation.

Most educators involved in agroecology and sustainable agriculture feel that a thoughtful analysis of the economic implications and long-term impacts of alternative systems will help us sort out the complexities of resource use and environmental impacts that result from alternative agricultural and food processing systems. A broad evaluation of productivity, economic return, environmental impact, and social equity that quantifies the multiple consequences of alternative food systems can help students and researchers ask the relevant questions, interpret the results, and apply this information to design of productive and sustainable agroecosystems. Moreover, it will be impossible to make substantial progress unless we also recognize the importance of literature, philosophy, ethics, and other reflections of culture that help explain the uniqueness of place, including people and resources. To introduce this concept to a broader student audience, we urge continued dialogue on both the definition of agroecology and the development of innovative approaches to resource use and food systems for the future in our agricultural colleges and universities world wide.

Indian agriculture in a way may be regarded as organic because majority of its cultivated area is under rainfed cultivation and only about 38% of cultivated land is under irrigation and in rainfed areas there is little or no use of fertilizers and other agriculture chemicals on account of risks associated, resource poor farmers and smaller land holdings.

Fertilizers are used only in 86.8% of the gross irrigated area and 42.7% of the un-irrigated area in the country (FAI, 2003).

Further, in these risk prone fragile ecosystems, the principal crops are pulses and oilseeds for which conventionally little or no fertilizer and other agriculture chemicals are used. So almost 62% of such areas of the country may be regarded as akin to organic farming (Marwaha and Jat, 2004).

Thus, the scope for practice of sustainable agriculture in India lies in delineation of prominent areas and identifying the crops suited for cultivation under low external input.

Among the prominent areas where sustainable farming can be easily adopted are fragile ecosystems like rainfed regions, hilly areas and north eastern states where fertilizer consumption is less than 25 kg/ha/year. In these areas awareness has to be created regarding the use of super-compost, vermicompost, biofertilisers and crop residues.

Large number of crops *viz.*, tea, coffee, cashew nuts and spices, coriander, turmeric may be easily brought under organic cultivation. Food grains like wheat, basmati and non-basmati rice, commercial crops like cotton, sugarcane, oilseeds like mustard, groundnut, sunflower, soybean etc., and certain medicinal and aromatic crops have been included in the list of crops having large potential for export.

In other words, promoting of organic cultivation should be done in selected regions and for selected crops to achieve sustainability in production. It would be desirable to delineate areas or rather demarcate the entire country into various organic zones which will help in formulation and adoption of technology package.

Significance / Importance

Modern crop production technology has considerably raised output but has created problems of land degradation, pesticide residues in farm produce, gene erosion, atmospheric and water pollutions. The natural resource base is degraded and diminished.

- The history of world reveals that great civilisations flourished along irrigation sources and mismanagement of these resources saw the extinction of these civilizations.
- With exploding population and rapid depletion and degradation of the natural resource base, sustainable

agriculture has assumed very great significance. The task of meeting the needs of present generation without eroding the ecological assets of the future generations is receiving top priority by environmental planners.

Historically, farming played an important role in human development. World population continues to grow. According to recent United Nations population projections, the world population will grow from 5.7 billion in 1995 to 9.4 billion in 2050, 10.4 billion in 2100, and 10.8 billion by 2150, and will stabilize at slightly under 11 billion around 2200. The rate of population increase is especially high in many developing countries. In these countries, the population factor, combined with rapid industrialization, poverty, political instability, and large food imports and debt burden, make long-term food security especially urgent.

First of all, why is there such an interest in sustainable agriculture? Eating food is necessary for survival, and most people desire maintaining a secure supply of food. Currently, the world's agricultural output exceeds the dietary needs of all humans on this planet. Widespread hunger in parts of the world are due mainly to the failure of political and economic systems to distribute the food.

So if enough food is being currently produced in the world, what is all this concern about "sustainable agriculture" ? The major concern is that while we may be able to feed the world today, future generations may not have adequate food production to satisfy their dietary needs. We know that some farming and land use techniques are not sustainable. For example, the Middle East used to be known as the "Fertile Crescent," but now much of that land is desert. Poor land use planning and unsustainable agricultural practices has lead to desertification and declining soil fertility in that region. Farmers and environmentalist alike are interested in creating cropping systems that have the potential to produce similar yields year after year with no decline in fertility.

There are many advantages of sustainable agriculture:

- Diversity of crops and products reduces the need for purchased fertilizers, helps in pest protection, and buffers the farm against unpredictable weather.
- Diversity also buffers the farm against changes in prices for one or a small number of commodities and increasing prices of production inputs.
- Crops raised by sustainable practices have more tolerance to stress conditions, although those under conventional management may do better under ideal growing conditions.
- There is increased effectiveness of sustainable systems compared to gradually reduced effectiveness of conventional systems – increasing pesticide resistance in weeds and insects, and increasing needs and costs for conventional fertilizers are seen as limitations to current standard recommendations and practices.
- "Sustainable agriculture is a model of social and economic organization based on an equitable and participatory vision of development which recognizes the environment and natural resources as the foundation of economic activity. Agriculture is sustainable when it is ecologically sound, economically viable, socially just, culturally appropriate and based on a holistic scientific approach.
- "Sustainable agriculture preserves biodiversity, maintains soil fertility and water purity, conserves and improves the chemical, physical and biological qualities of the soil, recycles natural resources and conserves energy. Sustainable agriculture produces diverse forms of high quality foods, fibers and medicines.

- "Sustainable agriculture uses locally available renewable resources, appropriate and affordable technologies and minimizes the use of external and purchased inputs, thereby increasing local independence and self sufficiency and insuring a source of stable income for peasants, family and small farmers and rural communities. This allows more people to stay on the land, strengthens rural communities and integrates humans with their environment.
- "Sustainable agriculture respects the ecological principles of diversity and interdependence and uses the insights of modern science to improve rather than displace the traditional wisdom accumulated over centuries by innumerable farmers around the world."
- "Sustainable agriculture does not refer to a prescribed set of practices. Instead, it challenges producers to think about the long-term implications of practices and the broad interactions and dynamics of agricultural systems. It also invites consumers to get more involved in agriculture by learning more about and becoming active participants in their food systems. A key goal is to understand agriculture from an ecological perspective—in terms of nutrient and energy dynamics, and interactions among plants, animals, insects and other organisms in agroecosystems —then balance it with profit, community and consumer needs."

Definitions of Sustainable Agriculture

The word "sustain" is derived from the Latin word *sustinere* (*sus-*, from below and *tinere*, to hold), to keep in existence or maintain, implies long-term support or permanence. As it pertains to agriculture, sustainable describes farming systems that are "capable of maintaining their productivity and usefulness to society indefinitely. Such systems must be resource-conserving, socially supportive, commercially competitive, and environmentally sound."

The term 'sustainable' implies a time dimension and the capacity of a farming system to endure indefinitely (Lockerer, 1988)

According to dictionary meaning "Sustainability" refers to keeping an effort continuously, the ability to last out & keep from falling.

In the context of agriculture, sustainability basically refers to the capacity to remain productive while maintaining the resource base.

Rutton (1988) pointed out that the concept of sustainability should serve as a guide to agricultural practice and it must also include the use of technology & practices that both sustainable and enhance productivity to meet the increasing food demands.

According to the dictionary meaning (Oxford English dictionary) the word sustainable is defined as the ability to maintain without deteriorating through continuos efforts.

"A systems approach is essential to understand sustainability. The system is envisioned in its broadest sense, from the individual farm, to the local ecosystem, and to communities affected by this farming system both locally and globally. A systems approach gives us the tool to explore the interconnections between farming and other aspects of our environment".

Environmental sustainability implies the following

- meeting the basic needs of all peoples, and giving this priority over meeting the needs of a few.
- keeping population densities, if possible, below the carrying capacity of the region.

- adjusting consumption patterns and the design and management of systems to permit the renewal of renewable resources.
- conserving, recycling, and establishing priorities for the use of nonrenewable resources
- keeping environmental impact below the level required to allow the systems affected to recover and continue to evolve.

"An environmentally sustainable agriculture is one that is compatible with and supportive of the above criteria" (Hill, 1992).

Dr. Hill further explains "To help recognize these real issues I distinguish between shallow (short-term, symbolic) and deep (long-term, fundamental) sustainability. Shallow sustainability focuses on efficiency and substitution strategies with respect to the use of resources. It usually accepts the predominant goals within society without question, and aims to solve problems by means of curative solution. Deep sustainability, in contrast, re-evaluates goals in relation to higher values and redesigns the systems involved in achieving these goals to that this can be done within ecological limits".

Sustainable agriculture is "a way of practicing agriculture which seeks to optimize skills and technology to achieve long-term stability of the agricultural enterprise, environmental protection, and consumer safety. It is achieved through management strategies which help the producer select hybrids and varieties, soil conserving cultural practices, soil fertility programs, and pest management programs. The goal of sustainable agriculture is to minimize adverse impacts to the immediate and off-farm environments while providing a sustained level of production and profit. Sound resource conservation is an integral part of the means to achieve sustainable agriculture" [USDA, 1999).

"Sustainable agriculture: A whole-systems approach to food, feed, and other fiber production that balances environmental soundness, social equity, and economic viability among all sectors of the public, including international and intergenerational peoples. Inherent in this definition is the idea that sustainability must be extended not only globally, but indefinitely in time, and to all living organisms including humans.

"Sustainable agriculture is a philosophy based on human goals and on understanding the long-term impact of our activities on the environment and on other species. Use of this philosophy guides our application of prior experience and the latest scientific advances to create integrated, resource-conserving, equitable farming systems. These systems reduce environmental degradation, maintain agricultural productivity, promote economic viability in both the short and long term, and maintain stable rural communities and quality of life" (Francis and Youngberg, 1990).

"Sustainable agriculture does not mean a return to either the low yields or poor farmers that characterized the 19th century. Rather, sustainability builds on current agricultural achievements, adopting a sophisticated approach that can maintain high yields and farm profits without undermining the resources on which agriculture depends."

An ecological definition of sustainable agriculture

Sustainable agriculture

A whole-systems approach to food, feed, and fiber production that balances environmental soundness, social equity, and economic viability among all sectors of the public, including international and intergenerational people. Inherent in this definition is the idea that sustainability must be extended not only globally but indefinitely in time, and to all living organisms including humans.

One modern definition is as follows

Sustainable agriculture is the use of farming systems and practices which maintain or enhance:

- the economic viability of agricultural production;
- the natural resource base; and
- other ecosystems which are influenced by agricultural activities.

Sustainability = Productivity + Resource Conservation

Sustainable agriculture is also known as ecofarming as ecological balance is given importance, natural farming or permaculture or organic farming as organic matter is the main source for nutrient management.

But some scientists consider that it is a misconception to think that sustainable agriculture is farming without chemical inputs. It is considered to be as integrated, low input and highly productive farming system.

As mentioned in the Integrated Farm Management Program of the 1990 Farm Bill as an example, sustainable farming is creative farming, and being creative means using the tools you have to work with. The IFM was an excellent example of how we can be creative, be sustainable and at the same time have the government help us make the changes to sustainable agriculture.

In defining sustainable agriculture, Fernholz reviews several information sources and concludes that sustainable practices must include reduced use of non-renewable resources, greater reliance on resources already on the farm, and increased interaction between the farmer and the land. He has been closely involved with various associations in Minnesota as well as with the Rodale Institute on-farm Research Network.

Based on his work & presentations he has summarized some current thinking about farmers' concerns and needs:

- farmers want to be involved
- producers have experience they are willing to share
- information is greatly needed including review of older results
- communication among producers is essential
- people, too often, fail to understand how sustainable agriculture fits into today's realities.

According to him LISA is ecologically desirable

Stinner and House (1987) have defined the sustainable agriculture systems as these which are typically associated with lower inputs.

What is sustainable agriculture?

The idea of sustainable agriculture has been around a long time. Since the very first crop was sown and animal was penned, farmers have tried to ensure that their land produces a similar or increasing yield of products year after back-breaking year; recent attempts to popularise the concept build on this tradition.

"Sustainable agriculture" is both a term and a concept whose definition has varied a great deal. As articulated in the 1990 "Farm Bill" Food, Agriculture, Conservation, and Trade Act of 1990, P.L. 101-624, Title XVI, Subtitle A, Section 1603) sustainable agriculture means "an integrated system of plant and animal production practices having a site-specific application that will, over the long term: (a) satisfy human food and fiber needs; (b) enhance environmental quality and the natural resource base upon which the agricultural economy depends; (c) make the most efficient use of nonrenewable resources and on-farm resources and integrate, where appropriate, natural biological cycles and controls; (d) sustain the economic viability of farm operations; and (e) enhance the quality of life for farmers and society as a whole."

Other definitions of sustainable agriculture

Simple stated, sustainable agriculture is a form of agriculture aimed at meeting the needs of the present generation without endangering the resource base of the future generations.

In order to feed the burgeoning population more food has to be produced and this has to be done without degradation of the resource base.

Expanding agriculture to ecologically fragile areas means greater threat to environment.

According to edwards (1987)

"Sustainable agriculture is minimal dependence on synthetic fertilizers pesticides & antibiotics & more dependent on use of manures, crop rotation & minimum tillage.

Technical Advisory Committee (TAC, 1989) of the Consultative Group on International Agriculture Research (CGIAR) has defined sustainable agriculture as "Successful management of resources for agriculture to satisfy changing human needs while maintaining or enhancing the quality of the environment and conserving natural resources".

From agroecological perspective, sustainability may be defined as a measure of productivity over time per unit input of nonrenewable or a limiting resource, per unit amount of soil erosion, per unit decline in soil organic matter, per unit energy input from fossil fuel, per unit increase in the concentration in the ground water, per unit of economic cost and value added (Lal & Miller, 1999).

According to Lockeretz (1986) sustainable agriculture is a loosely defined term that encompasses a range of strategies for addressing a number of problems that effect worldwide. Such problems include loss of soil productivity from excessive erosion and associated plant nutrient losses; surface and ground water pollution from pesticides and fertilizers; impending shortages of non renewable resources; and low farm income from depressed commodity prices and high production costs.

The Agricultural Research Service (U.S.Department of Agriculture) defines sustainable agriculture as:

"Agriculture that for the foreseeable future will be productive, competitive and profitable, conserve natural resources, protect the environment, and enhance public health, food quality, and safety".

The U.S. National Research Council (NRC, 1989) described that the ultimate goal of sustainable agriculture is to develop farming systems that are productive and profitable, conserve the natural resource base, protect the environment, and enhance health safety.

It is also referred to as the use of resources to produce food and fiber in such a way that the natural resource base is not destroyed, and that the basic needs of producers and consumers can be met over the long-term (Smit and Smithers, 1994 in Yunlong and Smit, 1994).

Admittedly, sustainable agriculture has several definitions, but it has three basic values: it is ecologically sound, economically viable and socially just and humane (Huang, 1994). Its major components in terms of agricultural technology are:

a) Cultural practices and plant breeding;

b) Soil and water management;

c) Non-chemical pest and weed control;

d) Integrated plant animal production; and

e) Nutrient recycling

The idea of "sustainable living" draws on the earlier concept of "sustainable development." "Sustainable development" was first put forward in 1980 in the World Conservation Strategy, a document of the International Union for the Conservation of Nature (IUCN) an international joint private / governmental group based in Geneva, Switzerland. Subsequently "sustainable development" became known globally with the publication of *Our Common Future*, the report of the United Nations appointed Brundtland Commission. The Brundtland definition of sustainable development "development that meets the needs of the present generation without compromising the ability of future generations to meet their own needs" was formally embraced. When 170 countries in attendance at the United Nations Conference on Event of Derint (UNCED), the "Earth summit" in Rio, Brazil, endorsed Agenda 21, the sustainable development agreement.

The concept of sustainable development is an evolving one, and there are many definitions in the literature, Pezzey (1992) lists 27 definitions of sustainability and sustainable development. According to Brundtland Report (WCED 1987):

Sustainable development is development that meets the needs of the present without compromising the ability of future generations to meet their own needs.

In 1988, and on the basis of the Brundtland Commission definition of sustainable development, the FAO Council defined SARD as:

The management and conservation of the natural resource base, and the orientation of technological and institutional change so as to ensure the attainment and continued satisfaction of human needs for present and future generations. Such sustainable development (in the agriculture, forestry and fisheries sectors) conserves land, water, plant and animal genetic resources, is environmentally non-degrading, technically appropriate, economically viable and socially acceptable (FAO, 1989).

Sustainable agroecosystems (Gliessman, 1998)

- maintain their natural resource base
- rely on minimum artificial inputs from outside the farm system
- manage pests and diseases through internal regulating mechanisms
- recover from the disturbances caused by cultivation and harvest."

Sustainable farming practices commonly include

- crop rotations that mitigate weeds, disease, insect and other pest problems; provide alternative sources of soil nitrogen; reduce soil erosion; and reduce risk of water contamination by agricultural chemicals
- pest control strategies that are not harmful to natural systems, farmers, their neighbours, or consumers. This includes integrated pest management techniques that reduce the need for pesticides by practices such as scouting, use of resistant cultivars, timing of planting, and biological pest controls
- increased mechanical/biological weed control; more soil and water conservation practices; and strategic use of animal and green manures
- use of natural or synthetic inputs in a way that poses no significant hazard to man, animals, or the environment.

"This approach encompasses the whole farm, relying on the expertise of farmers, interdisciplinary teams of scientists, and specialists from the public and private sectors" (Connell, 1992).

Sustainability of our agricultural systems is of global concern today and many definitions of sustainable agriculture have become available. The five main components of these definitions are:

- production of enough food and fibre to meet the increasing and changing needs of the people;
- conservation of natural resources;
- maintaining the quality of environment;
- achieving community and gender equity; and
- avoidance of regional imbalances

Objectives

"Sustainable agriculture" was addressed by Congress in the 1990 "Farm Bill" [Food, Agriculture, Conservation, and Trade Act of 1990 (FACTA), Public Law 101-624, Title XVI, Subtitle A, Section 1603 (Government Printing Office, Washington, DC, 1990). Under that law, "the term sustainable agriculture means an integrated system of plant and animal production practices having a site-specific application that will, over the long term:

- satisfy human food and fiber needs
- enhance environmental quality and the natural resource base upon which the agricultural economy depends
- make the most efficient use of nonrenewable resources and on-farm resources and integrate, where appropriate, natural biological cycles and controls
- sustain the economic viability of farm operations
- enhance the quality of life for farmers and society as a whole."

It integrates three main goals, environmental health, economic profitability, and social and economic equity. These goals have been defined by a variety of philosophies, policies and practices, from the vision of farmers and consumers. Perspectives and approaches are very diverse, the following topics intend to help understanding what sustainable agriculture is.

Fred Kirschenmann provides a prescriptive approach to the conversion process, one that he suggests will take both time and creativity on the part of the manager. He cites the objectives used by the IFOAM (International Federation of Organic Agriculture Movements) as what need to be developed during a transition: fertility-building crop rotation, proper manure management, appropriate tillage / cultivation, and an environment that reduces pests (insects, weeds, diseases). As a sidelight, he mentions the over riding importance of crop rotations as providing a route to reaching several of these objectives.

Nonetheless, the principles and practices that lie behind these terms are essentially similar. The objectives of organic agriculture are been concisely expressed in the standard document of the International Federation of Organic Agriculture Movement (IFOAM) as follows:

- To produce food of high nutritional quality in sufficient quantity.
- To work with natural systems rather than seeking to dominate them
- To encourage and enhance the biological cycles within farming system involving microorganisms, soil flora and fauna, plants and animals.
- To maintain and increase the long term fertility of soils
- To use, as far as possible, renewable resources in locally organized agricultural systems.

- To work, as much as possible, within a closed system with regard to organic matter and nutrient elements.
- To give all livestock, conditions of life that allow them to perform all aspects of their innate behaviour.
- To avoid all forms of pollution that may result from agricultural techniques.
- To maintain the genetic diversity of the agricultural system and its surroundings, including the protection of plant and wildlife habitats.
- To allow agricultural producers an adequate return and satisfaction from their work including a safe working environment; and
- To consider the wider social and ecological impact of the farming system.

What is sustainable agriculture? Environmental Science, an introductory course book for environmental sciences, gives three criteria for sustainable agriculture. (i) It must feed the world's hungry today. (ii) It must feed the world's hungry tomorrow. (iii) It must prevent deterioration of soil & water.

This sounds good, but how is it achieved? In today's political environment, people are not concerned with just having enough food but with many other factors as well. Agricultural systems have to not only produce crops, but they must also produce food that is safe to eat, minimize ecological impacts, compete with other land uses, produce affordable foods, work within changing political & economic climates, and create products consumers are willing to buy. Within this framework establishing sustainable agricultural systems is much more difficult than it would first seem.

The design of sustainable agroecosystems (Pretty 1994, Vandermeer 1995)

Most people involved in the promotion of sustainable agriculture aim at creating a form of agriculture that maintains productivity in the long term by :

- Optimizing the use of locally available resources by combining the different components of the farm system, i.e. plants, animals, soil, water, climate and people, so that they complement each other and have the greatest possible synergetic effects;
- Reducing the use of off-farm, external and non-renewable inputs with the greatest potential to damage the environment or harm the health of farmers and consumers, and a more targeted use of the remaining inputs used with a view to minimizing variable costs;
- Relying mainly on resources within the agroecosystem by replacing external inputs with nutrient cycling, better conservation, and an expanded use of local resources;
- Improving the match between cropping patterns and the productive potential and environmental constraints of climate and landscape to ensure long-term sustainability of current production levels;
- Working to value and conserve biological diversity, both in the wild and in domesticated landscapes, and making optimal use of the biological and genetic potential of plant and animal species; and
- Taking full advantage of local knowledge and practices, including innovative approaches not yet fully understood by scientists although widely adopted by farmers. Agroecology provides the knowledge and methodology necessary for developing an agriculture that is on the one hand environmentally sound and

on the other hand highly productive, socially equitable and economically viable.

Sustainable agriculture must produce enough food and fibre to satisfy changing human needs while conserving natural resources, maintaining the quality of environment and ultimately leading to community and gender equity.

"Sustainable agriculture is a goal, not a fixed technology, but an ever-evolving approach to farming that changes as the body of knowledge grows about ecosystems and agriculture." Fernholz suggests that evaluating management should include profiles of variable costs/acre, pesticide usage, crop performance, and energy/capital/labour used on the farm. As an example, he suggests comparison of a farmer's own production profile with the state averages, or with a comparable "control group," that is relevant for a smaller area. He emphasizes the importance of setting realistic goals and examining where time and investment will give the greatest returns, consistent with the broader goals of the family.

❐❐❐

2 CONCEPT OF SUSTAINABLE AGRICULTURE

The concept of multiple dimensions of a sustainable agriculture is more broad which has to be viewed with awareness of potential technologies, fragile ecosystem and human interference to disrupt with limited resource availability. All these aspects have to be taken into account for proper understanding of sustainable agriculture.

The major aspects are,

1. Agriculture must be increasingly productive and efficient in resource use.
2. Biological processes within agriculture systems must be controlled from within (rather than by external inputs of pesticides) and
3. Nutrient cycles within the farm must be closed.

Within this frame work a sustainable model has to be developed. The proposed concept of sustainable agriculture which is best suited for developing and developed countries and is a viable model for the present and future is depicted schematically (Fig. 1).

In its purest form sustainable agriculture would be a complete reversion to subsistence agriculture relying on organic wastes and manures within farms without applying

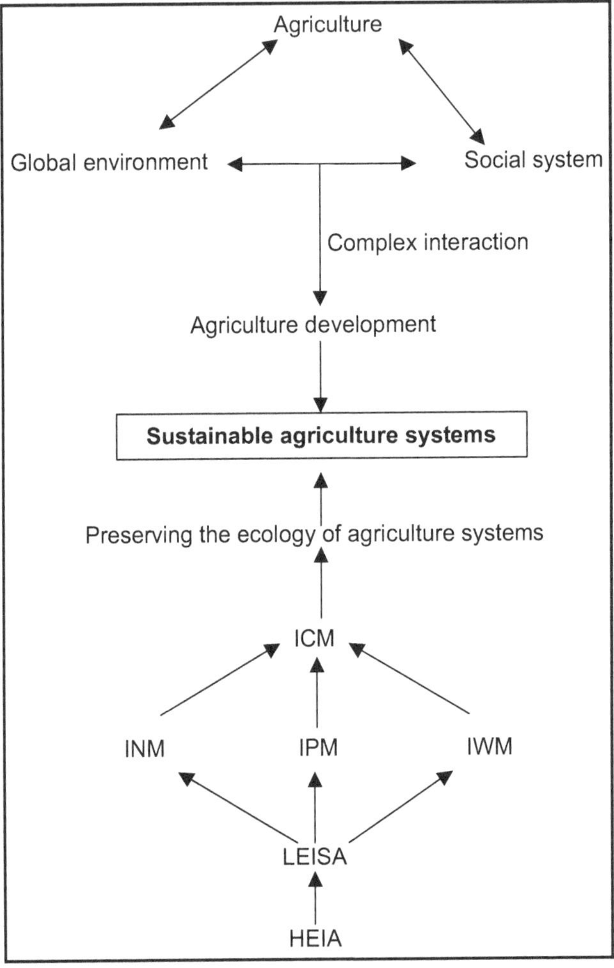

Fig.1. Present and Future concept of sustainable agriculture

external fertilizer or other chemicals. But this possibility seems to be risky with the world's current population growth and the total agriculture need of food and fibre. Under these circumstances the proposed concept should be looked into, because true sustainability will remain an ideal to be approached but not really obtainable. But striving for it as a critical target should continue without over-burdening the resource base. If natural resources such as soil, nutrients and water are used at a rate faster than they are replenished, then the farming system is unsustainable. Sustainability is also dependent on maintaining a high level of biodiversity, especially in the soil and the surrounding environment. Practically, farmers will adopt systems which will provide a living for themselves and their families. Therefore, economic and social issues as well as the productivity of land and the broader health of the environment have to be considered when working towards sustainable agriculture.

In order to fully understand the concept of sustainability we need to understand all aspects such as, environmental, economic and social sustainability i.e. how agriculture depends on certain resources and the environmental, economic and social implications needs to be analysed.

The economic component of sustainability depends upon a long term path of growth that leads to real growth incomes. The social component lies in the assumption that an acceptable quality of life can be maintained in the long term only if equal opportunities exist in the implementation of developmental options. The environmental component is related to the long term functional capacity and the productivity of life-preserving natural ecosystems. This integration of development goals is the special feature of the concept.

Technical Advisory Committee (TAC) of CGIAR in its paper "Sustainable Agriculture Production: Implications for International Agriculture Research", discusses sustainability as a dynamic concept that allows for changing needs of a steadily increasing global population. "Sustainable agriculture should involve the successful management of resources for agriculture to satisfy changing human needs while maintaining or enhancing the quality of the environment and conserving natural resources".

What is new about sustainable agriculture ?

The concept of sustainable really constitutes the acceptance of a broadened agenda for agricultural research. It represents a shift in balance, greater environmental concern & renewed importance given to biological and social interpretation. Historically, agricultural systems were viewed as a relatively linear interaction between inputs made to crops & soils, leading to a particular yield (Fig. 1). The management actions were focused at a field scale over a time period of a few crop cycles, with a single objective of yield maximization. Recently global concerns have broadened this view to include biodiversity, nutrient budgets, water quality and quantity forcing a more complex interactive perspective (Fig. 1).

What is new about the concept of sustainable agriculture is the application of biological principles to enhance stability, resource use efficiency and productivity. Such a situation has occurred in response to a growing awareness of the interdependence of agriculture, the environment and socio economic conditions.

Criteria of sustainability : Judging agriculture to be sustainable if it fulfills the following conditions (Gips, 1986)

1) *Ecologically sound:* Which means that quality of natural resources is maintained & the vitality of the entire agro-ecosystem from human beings, crops and animals to soil organisms is enhanced. This is best ensured when the soil is managed and the health of crops, animals & people is maintained through biological processes.

2) *Economically viable:* Which means that the farmers can produce enough for self sufficient income and gain sufficient returns. Economic viability is measured not only in terms of direct produce but also in terms of functions such as conserving resources and minimising risks also.

3) *Socially Just:* Which means that the resources and power are distributed in such a way that the basic needs of all members of society are met and their rights to land use, adequate capital, technical assistance are assured. All people should have opportunity to participate in the decision making.

4) *Humane:* Which means that all forms of life (plant, animal, human) are respected. The cultural and spiritual integrity of the society is preserved.

5) *Adaptable:* Which means that the rural communities are capable of adjusting to constantly changing conditions for farming, population growth, national policies and market fluctuations.

Classification of sustainable agriculture systems

While sustainable agriculture has become the umbrella under which, many of the alternative farming systems take shelter. It is however important to realize that sustainable agriculture is a long term goal and not a specific set of farming practices.

Sustainable agriculture is a philosophy based on human goals and on understanding the long-term impact of our activities on the environment and other species.

In this context, sustainable agriculture embraces all agriculture systems, striving to meet these criteria. However, from the point of view of better understanding of the concept, sustainable agriculture system can be classified as shown below.

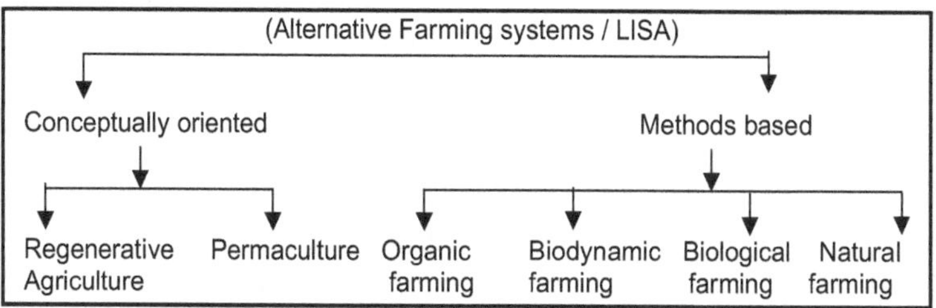

Classification

Sustainable Agriculture is an umbrella term which encompasses all other synonymous terms *viz.,* organic farming, biodynamic farming, eco-farming, regenerative agriculture, LEISA.

Sustainable agriculture integrates three main goals—environmental health, economic profitability, and social and economic equity. A variety of philosophies, policies and practices have contributed to these goals. People in many different capacities, from farmers to consumers, have shared this vision and contributed to it. Despite the diversity of people and perspectives, the following themes commonly weave through definitions of sustainable agriculture.

Sustainability rests on the principle that we must meet the needs of the present without compromising the ability of future generations to meet their own needs. Therefore, *stewardship of both natural and human resources* is of prime importance. Stewardship of human resources includes consideration of social responsibilities such as working and living conditions of labourers, the needs of rural communities, and consumer health and safety both in the present and the future. Stewardship of land and natural resources involves maintaining or enhancing this vital resource base for the long term.

A system perspective is essential to understanding sustainability. The system is envisioned in its broadest sense, from the individual farms, to the local ecosystem, *and* to communities affected by this farming system both locally and globally. An emphasis on the system allows a larger and more thorough view of the consequences of farming practices on both human communities and the environment. A systems approach gives us the tools to explore the interconnections between farming and other aspects of our environment.

A systems approach also implies interdisciplinary efforts in research and education. This requires not only the input of researchers from various disciplines, but also farmers, farm workers, consumers, policy makers and others.

Making the transition to sustainable agriculture is a process. For farmers, the transition to sustainable agriculture normally requires a series of small, realistic steps. Family economics and personal goals influence how fast or how far participants can go in the transition. It is important to realize that each small decision can make a difference and contribute to advancing the entire system further on the "sustainable agriculture continuum." The key to moving forward is the will to take the next step.

Finally, it is important to point out that reaching towards the goal of sustainable agriculture is the responsibility of all participants in the system, including farmers, labourers, policymakers, researchers, retailers, and consumers. Each group has its own part to play, its own unique contribution to make to strengthen the sustainable agriculture community.

The remainder of this document considers specific strategies for realizing these broad themes or goals. The strategies are grouped into three separate though related areas of concern: farming and natural resources, plant and animal production practices, and the economic, social and political context. They represent a range of potential ideas for individuals committed to interpreting the vision of sustainable agriculture within their own circumstances.

In addition to these methods based approaches to sustainable farming, regenerative agriculture and permaculture are widely recognized in the U.S. and abroad. However, these latter systems, like sustainable agriculture, are more conceptually oriented than methods-based.

The concept of sustainable agriculture is a relatively recent response to the decline in the quality of the natural resource base associated with modern agriculture (McIsaac and Edwards, 1994). Today, the question of agricultural production has evolved from a purely technical one to a more complex one characterized by social, cultural, political and economic dimensions. The concept of sustainability although controversial and diffuse due to existing conflicting definitions and interpretations of its meaning, is useful because it captures a set of concerns about agriculture which is conceived as the result of the co-evolution of socio-economic and natural systems (Reijntjes *et al.,* 1992). A wider understanding of the agricultural context requires the study between agriculture, the global environment and social systems given that agricultural development results from the complex interaction of a multitude of factors. It is through this deeper understanding of the ecology of agricultural systems that doors will open to new management options more in tune with the objectives of a truly sustainable agriculture. The sustainability concept has prompted much discussion and has promoted the need to propose major adjustments in conventional agriculture to make it more environmentally, socially and economically viable and compatible. Several possible solutions to the environmental problems created by capital and technology intensive farming systems have been proposed and research is currently in progress to evaluate alternative systems (Gliessman, 1998). the main focus lies on the reduction or elimination of agrochemical inputs through changes in management to assure adequate plant nutrition and plant protection through organic nutrient sources and integrated pest management, respectively. The prevalent philosophy is that the pests, nutrient deficiencies or other factors are the cause of low productivity, as opposed to the view that pests or nutrients only become limiting if conditions in the agro-ecosystem are not in equilibrium (Carrol *et al.,* 1990). For this reason, there still prevails a narrow view that specific causes affect productivity, and overcoming the limiting factor via new technologies, continues to be the main goal. This view has diverted agriculturists from realizing that limiting factors only represent symptoms of a more systemic disease inherent to imbalanced within the agro-ecosystem and from an appreciation of the context and complexity of agro-ecological processes, thus underestimating the root causes of agricultural limitations (Altieri *et al.*, 1993).

'Sustainability' is one of the buzz-words of our times. It has been attached to any number of social and economic pursuits: we want sustainable economic growth, sustainable development, sustainable forestry, a sustainable population, sustainable cities, and so on. Increasingly, we also talk about the need for "sustainable agriculture". We need to define it in terms that farmers can understand and to develop ways in which they can measure it. Above all, perhaps, we need to ensure that the concept is useful to farmers.

❑❑❑

3 HISTORY AND DEVELOPMENT OF SUSTAINABLE AGRICULTURE

Development of Agriculture in the World and in India

Agricultural ecosystems will increasingly be called on to deliver food and fiber for a demanding and expanding global population. Until recently, increased production relied on expansion of the amount of land under cultivation and irrigation. Since World War II, production advances have relied on genetics and plant breeding, advances in fertilizer and pesticide use and improved cultural management (Matson *et al.,* 1997). Since 1950, agriculture worldwide has expanded cereal production by an average of 2% per annum while population has increased by 1.9% per year (Borlaug and Dowswell, 1994).

Modern science-based agriculture had its beginnings in the late 19th century with improved traction and tillage equipments. However, it was not until after World War II that the research findings that led to the modern-day productive systems in the United States and later in European and other developed countries. The expansion was directly related to the great demand for food to support the Allied war effort. Knowledge bases developed in plant genetics, physiology and pathology, along with a ten-fold increase in fertilizer use, advances in seed technology, weed control, irrigation and cultural methods allowed the development of the systems that dominate agriculture in many of the countries today. Corn or maize responded dramatically to the advances in production technology. Maize yield in the United States has increased more than 300 per cent over the past 50 years and continues to rise. Maize production technology has been delivered worldwide (Borlaug and Dowswell, 1994).

The availability of land and productivity of soil limit food production; however, it would have been difficult to support the present population if some significant advances in agriculture and food production had not occurred.

Three major scientific steps changed the whole process of food production almost all over the world.

1. Mendel's laws of inheritance postulated in 1866 were rediscovered in 1900. Plant hybridization gained momentum. The concept of gene emerged which now is the foundation of genetic engineering and biotechnology. However, the most dramatic effect of the use of genetics in crop yield improvement occurred through the mechanism of heterosis and hybrid vigour. The higher yields of maize obtained through this mechanism made crop improvement attractive to entrepreneurs and led to commercialization. This also brought into prominence and focus the importance of germplasm and biodiversity.
2. Recognization of nutrient requirement and fertilizer use. It was shown by chemists that plants contained many elements in varying quantities. These nutrients must be obtained from soil. This led to the concept of fertilizers and its usage. The chemical process of ammonia production was a major step in this respect which has revolutionised the whole fertilizer industry and world agriculture.

Realization that plants could mobilize nutrients from the soil only when it was adequately wet and the concept of irrigation which was existed in vague helped in the development of major irrigation projects world over.

There were several famines in India in the nineteenth century. India experienced unprecedented droughts and the consequent famine in 1899 and 1900 which led the then government to establish the Irrigation Commission in 1901. One of the major recommendations of this commission was to develop irrigation to reduce the impact of drought and the chances of famine through increased production. More area was brought under cultivation, but there really was no significant impact on the productivity of three major crops *viz.,*wheat, mustard and gram (chickpeas). Wheat yield did not change until 1965 and that of rape-seed and mustard until 1985. However, a dramatic change in productivity of wheat started from 1970 and there is a slowing down of productivity of wheat during recent years.

3. The productivity and production of crops has often been reduced by diseases and insect pests. In mid-1930s, pesticides, particularly for insect pest control, were discovered and quickly became a component of agricultural practices. Some of the worst famines such as the one in Ireland and Bengal in India were triggered by plant diseases. Therefore, overcoming plant diseases became a major objective around the world. The genetic control of wheat rust in India was achieved earlier than the arrival of Mexican dwarf (Rht-1 and rht-2 genes-based dwarfism along with other traits) wheats to India. A detailed study of the genetics of rust resistance shows that most Indian wheat have an important trait.

However, the insect pest problems continue to cause losses in crop productivity. Recent advances in technology such as genetic engineering have led to significant increase in yield of many field crops. Genetic engineering resulting in insect-resistant transgenic plants and crops is a major effort during recent years. Consequently 38% of the transgenic crops are for making plants resistant against diseases and insect pests. To feed the worlds rising population with the limited resources is at the top of the agenda of all the countries and hence sustainable agriculture has acquired a different perception and meaning to different countries.

Not only in Africa has the persistence of low yield farming created threats to the rural environment but also in South Asia as well. Large dryland areas are suffering from degradation, deforestation and habitat loss due to an extended use of traditional farming practices under population pressures. The Hindu Kush-Himalaya region in South Asia, stretching from Pakistan across northern India, Nepal, Bhutan, and China, is generally too hilly and too remote for irrigated farming and highly prone to rapid erosion if the trees are cut for rainfed farming. Historically, these upper watershed areas were under forest cover, but over time and under population pressure they have been exploited for various kinds of low yield agriculture, and now only 19 per cent of the area remains under forest which is a great concern for sustainable agriculture.

One must consider the conditions that were prevailing in the late 1940s and early 1950s when India and other colonised countries witnessed a withdrawal of colonial powers. India suffered one of the worst famines, which claimed about five million people. Another five million died during the transfer of power on the eve of independence. Agriculture was in serious trouble. Shortage of food, and not maldistribution, was cited as the reason for the deaths. The United States enacted PL480 for free supply of food to

India and other starving countries. Actually the food was not supplied free; the importing countries were required to pay the cost but the amount remained deposited in these countries. The Ford Foundation and umpteen other charity organisations were given the deposited amount to carry out welfare programmes.

The integral sustainable agriculture concept developed in India, based on thousands of years of farming with the specific knowledge of local and regional climatic conditions, was in shambles. The British changed the crop pattern in India to suit the needs of Britain's Industrial Revolution. Fine cotton, with which the finest Dacca muslin was woven, was replaced by coarse cotton to meet the needs of Manchester and Lancashire. We were depending on doles. So the Green Revolution and the Japanese method of rice cultivation were introduced.

In India, much more of this kind of damage would have occurred without the yield-boosting impact of the original Green Revolution. Thanks to the introduction of high yield farming into India in the 1960s and 1970s. The area of land planted to crops did not have to expand as quickly to boost total food production in pace with population growth. In 1964, India produced 12 million tonnes of wheat on 14 million hectares of land. By 1993, India was producing 57 million tonnes of wheat on 24 million hectares of land.

Unfortunately, not all farmers in India work on lands that are suited to the high yield varieties. Dryland farmers still represent 40 percent of India's entire population, and among these farmers the average grain yields are only one third as high as on the well-watered lands that were able to use the new green revolution seeds. For these people poverty and hunger persists, and cropland area expansion into forests persists, because an affordable way to increase the productivity of land already under the plow has not yet been found.

The hybrid seeds that were dumped on India required massive inputs of synthetic fertilizers and toxic pesticides. No parasite existed in the local food chain for the pests that came along with the imported seeds and food. The American bollworm is one such gift of American food imports. The government introduced subsidies and incentives linked to the use of hybrid seeds, fertilizers and pesticides. Records show that the average yield of rice in Chingleput district [now Kanchipuram] in Tamil Nadu was 3,500 kg a hectare against 1,590 kg a hectare a couple of years ago after the changeover to the Green Revolution package. The Green Revolution changed the mindset of the people, particularly the farmers. Whatever the government said was good and worth copying. Subsidies and incentives support this thinking.

In Central America as well, the continued use of traditional low yield farming practices by poor communities experiencing rapid population growth has helped devastate the rural environment. Wealthy commercial farmers on good land in Central America are harming the environment by over-irrigating and spraying too many chemicals, but the vast majority of farmers in the region are impoverished and confined to poor lands where they struggle with traditional methods to produce enough corn and beans to keep their growing families fed. As population continues to increase and these poor lands become exhausted, farmers cut more trees, move higher up the hillside, and invade more wildlife habitat. Since 1961, Mexico has lost about 36 percent of its total forested area, and more than half of the soil area in Mexico is either eroded or undergoing accelerated erosion. In large parts of rural Mexico where impoverished peasant farmers have been struggling with traditional techniques to produce more on dry hillside lands, all the trees have now been cut, the thin soil has washed away, and the terrain resembles a lifeless moonscape. In the Mixteca region, 70 percent of the potentially arable land is no longer able to grow crops due to soil erosion.

The worst aspect of this kind of rural environmental damage from low yield farming is that it does greatest harm to the poor farmers themselves. In contrast to environmental damage from high yield farming (which frequently harms water users downstream from farms), this environmental damage from low yield farming is first of all a problem for the poor farmers themselves, since it usually results in a further reduction of productivity on their own farming lands. When high yield farmers in wealthy countries do environmental damage, their own high yields and their own commercial prosperity is seldom threatened, but when low yield farmers in poor countries do damage - depleting soil nutrients, cutting trees, ruining watersheds, over-grazing arid lands - it is their own resource base that suffers, reducing their own production prospects still more.

Development of Sustainable Agriculture

In the early 1900s, several works focused on the broader, non-simplistic aspects of agriculture and their interrelationships.

Elliot (1907) wrote of the complexities of pasture mixes and their importance to soil fertility in rotations.

King wrote two books *viz.*, Farmers of forty centuries (1911) and soil management (1914), in both the books he stressed on the complexity of integration in the then highly productive, traditional systems of Asia.

Biodynamic Movement

In Germany, Rudolf Steiner <http://en.wikipedia.org/wiki/Rudolf_Steiner>'s *Spiritual Foundations for the Renewal of Agriculture*, published in 1924, led to the popularization of biodynamic agriculture <http://en.wikipedia.org/wiki/Biodynamic_agriculture>, one of the first organic farming systems.

Since 1920's, biodynamic farmers have developed the execution of principles of diversification, recycling, avoiding chemicals, decentralized production & distribution, thus biodynamic movement started with a series of lectures given by Rudolf Steiner, the founder of anthroposophy, in 1924. Biodynamic practices includes stimulation and regulation of complex life process by biodynamic preparations. It also includes the consideration of cosmic and terrestrial forces on biological organisms.

Humus Farming

This concept provided the foundation of biological agriculture that emerged from the 1930s through the 1960s. As early as in 1855 Brown, wrote "The Field Book of Manures", followed by Roberts (1907), Eletcher (1907) and Waksman (1936) wrote on humus oriented soil fertility.

Later on advanced techniques of compost making added significantly to the emerging thought of humus farming. In 1943, Sir Albert Howard's book 'An Agricultural Testament', became a milestone. Indore method of composting was widely published. This work influenced the soil society work in England as well as the writing of J.I. Rodale in the United States. The principles of composting and compost use was popularized which paved way for the organic farming movement that followed.

Organic Movement

The organic movement <http://en.wikipedia.org/wiki/Organic_movement> began as a reaction of insiders (agricultural scientists and farmers) against the industrialization of agriculture. For some time it remained below the awareness of the food buyer. As the contrasts between organics and the new conventional agriculture grew, so did public awareness of organic farming. This led to a distinct organic market and eventually, a grassroots consumer cause.

In 1939 Lady Eve Balfour <http://en.wikipedia.org/wiki/Lady_Eve_Balfour>, influenced by Sir Howard's work, launched the first scientific, side-by-side comparison of organic and conventional farming in England. Called as the Haughley Experiment <http://en.wikipedia.org/wiki/Haughley_Experiment>, it was documented by Lady Balfour in her book, The Living Soil <http://en.wikipedia.org/wiki/The_Living_Soil>. Its influence led to the formation of the Soil Association <http://en.wikipedia.org/wiki/Soil_Association>, a key international organic advocacy group.

While some indigenous cultures had been farming organically for centuries, organic agriculture began to develop consciously in Central Europe and India in the early twentieth century as a reaction to industrialization. The British botanist, Sir Albert Howard <http://en.wikipedia.org/wiki/Sir_Albert_Howard> often called "the father of modern organic agriculture" studied traditional farming practices in Bengal <http://en. wikipedia.org/wiki/Bengal>, India <http://en.wikipedia. org/wiki/India>. In 1940, he came to regard such practices as superior to modern agricultural science and recorded them in his book, *An Agricultural Testament <http://en.wikipedia.org/wiki/An_Agricultural_Testament>* which provided the basic tenet that led to organic, biological and eventually to regenerative farming.

The ideas of an integrated, decentralized, chemical free agriculture were advocated by Northburn (1940). He was the first to use the word organic in reference to the entire philosophy and practice.

The first use of the term *organic farming* is usually credited to Lord Northbourne <http://en.wikipedia.org/wiki/Baron_Northbourne>, in his book, *Look to the Land <http://en.wikipedia.org/w/index.php?title=Look_to_the_Land&action=edit>* (1940), wherein he described a holistic, ecologically balanced approach to farming.

The term organic, as a descriptor for certain sustainable agriculture systems, appears to have been first widely used by Lord Northbourn (1940) in his book "Look to the Land". Northbourn used the term to describe farming systems that focused on the farm as a dynamic, living, balanced, organic whole, or an organism. The term, thus, had broader meaning than just the use of living materials to achieve farming objectives, a restrictive definition that is often erroneously implied today. Its original meaning, then, is much closer to the origin of the term organic used in organic chemistry, the study of the chemistry of organisms. Unfortunately, many scientists continue to equate the term with the present-day meaning of organic chemistry, the study of carbon-containing compounds.

As early as in 1940 there was serious thinking to reconsider the existing agricultural practices. Barlow (1942) was specially critical and of the impact of agriculture on soil degradation and reduction in diversity through specialization.

The organic movement was furthered by Lady Eve Balfour's "The Living Soil" (1943) and Faulkner's, "Plowman's Folly" (1943).

In 1945, J.I. Rodale's 'Pay Dirt' became a rallying point that carried the organic movement.

The term organic was first widely used in the USA by J.I. Rodale, founder of Rodale Press, in the 1950s. Rodale was both the popularizer of the term organic (and by implication notions of sustainability), but also, in the scientific community, the inspiration for the denigration of the term. Rodale failed to convince scientists of the validity of his approach because of his reliance on what were perceived to be outrageous unscientific claims of organic farming's benefits. This was unfortunate as a number of scientists in the USA and Europe were investigating and promoting sustainability in agriculture at the time, most notably Sir Albert Howard and William Albrecht. The scientific and governmental fascination with using agrichemicals, monoculture, and specialized equipment for food production severely constrained professional interest in questions of sustainability.

Louis Bromfield also contributed significantly to organic movement and emphasized on organic farms on which people, crops and livestock were integrated in a living system.

Technological advances during World War II <http://en.wikipedia.org/wiki/World_War_II> spurred on post-war innovation in all aspects of agriculture, resulting in such advances as large-scale irrigation, fertilization, and the use of pesticides <http://en.wikipedia.org/wiki/Pesticide>. Ammonium nitrate <http://en.wikipedia.org/wiki/Ammonium_nitrate>, used in munitions, became an abundantly cheap source of nitrogen. DDT <http://en.wikipedia.org/wiki/DDT>, originally developed by the military to control disease-carrying insects among troops, was applied to crops, launching the era of widespread pesticide <http://en.wikipedia.org/wiki/Pesticide> use.

Several other authors stressed on the increasing environmental harm and resource degradation brought by 'modern' farming methods. All of them repeatedly advocated the holistic approach to agriculture.

The Green Revolution

Inspite of massive and unprecedented increases in population since the 1950s and in the face of predictions of (and actual instances of) starvations in Asia, most countries including India, Bangaldesh, China, the Philippines, Indonasia etc., have achieved self sufficiency and even food surpluses.

Advances in biochemistry (nitrogen fertilizer) and engineering (the internal combustion engine) in the early

20th century led to profound changes in farming. Research in plant breeding <http://en.wikipedia.org/wiki/Plant_breeding> produced hybrid seeds <http://en.wikipedia.org/wiki/Hybrid_seed>. Fields grew in size and cropping became specialized to make efficient use of machinery and reap the benefits of the so-called "green revolution <http://en.wikipedia.org/wiki/Green_revolution>."

The approach has worked best, however, in areas with good soil and water resources where response to technology application and inputs were high.

In the 1970s, global movements concerned with the environment championed organic farming. As the distinction between organic and conventional food became clearer, one goal of the organic movement was to encourage consumption of locally grown food <http://en.wikipedia.org/wiki/Local_food>, which was promoted through slogans like "Know Your Farmer, Know Your Food". In 1972, the International Federation of Organic Agriculture Movements (IFOAM), was founded in Versailles <http://en.wikipedia.org/wiki/Versailles>, France. IFOAM was dedicated to the diffusion of information on the principles and practices of organic agriculture across national and linguistic boundaries.

Evolution of Sustainable Agriculture

In the 1960s and 70s, a growing environmental agriculture movement evolved in response to increasing soil erosion, pesticide use, and groundwater contamination. Simultaneously, economic conditions for farmers were becoming more stressful and the number of family farms declined.

In 1980 Wes Jackson of The Land Institute in Salina, KS, began using the term "sustainable agriculture" to describe an alternative system of agriculture based upon resource conservation and quality of rural life. Through the lobbying efforts of several nonprofit farming organizations, Congress passed legislation in the 1985 Farm Bill that mandated implementation of a low-input sustainable agriculture program by the Department of Agriculture.

In 1988, USDA initiated the Low-Input Sustainable Agriculture research and education program, or LISA. In 1991 the name of this program was changed to the Sustainable Agriculture Research and Education Program, or SARE. Funds made available through the LISA/SARE programs have resulted in significant additions to land grant research and extension programs in the later five years.

While sustainable agriculture has become the umbrella under which many of the above-mentioned alternative farming systems fall, it is important to note that sustainable agriculture is really a long-term goal, not a specific set of farming practices. In Temperate Zones sustainable agriculture was defined as:

Sustainable Agriculture is a philosophy based on human goals and on understanding the long-term impact of our activities on the environment and on other species. Use of this philosophy guides our application of prior experience and the latest scientific advances to create integrated, resource-conserving, equitable farming systems. These systems reduce environmental degradation, maintain agricultural productivity, promote economic viability in both the short and long term and maintain stable rural communities and quality of life.

Three indicators that appear most frequently in a definition of sustainable agriculture are:

- Environmentally sound
- Economically viable
- Socially acceptable

In this context, sustainable agriculture embraces all agricultural systems striving to meet these criteria. Many aspects of modern conventional agriculture are included in sustainable agriculture, just as are many aspects of alternative farming systems.

One aspect of modern agriculture receiving a lot of attention in the sustainable agriculture discussion is the use of chemical inputs to supply fertility and pest control. While agriculture chemicals will continue to play an important role in American agriculture, many farmers are looking at alternatives due to environmental, economical, or regulatory reasons. In a transition to farming systems more reliant on biological methods of production, low-input farming serves as an intermediary step.

Proposals for sustainable agriculture development goes back to as far the "Limits to growth" report from Rome club in 1972. This report predicted soil pollution and the soil depletion caused by the intensive modern agriculture. This report emphasized that developing countries should not unconditionally adopt intensive agriculture and the dangers to the event from the over use of fertilizer in developed nations, and recommended the use of organization fertilizer and biological fixing of nitrogen.

IFOAM, 1972: International federation of organic agriculture movement was founded in Versailles, France, it has today 670 member organizations and institutions over 120 countries including India. The main thrust of IFOAM is to define concept of organic farming through their basic stands, harmonize certification programme, participation in UN, FAO and WHO and communication through seminar, magazines etc.

In 1985, the US congress passed the Agricultural productivity act as part of food security act, public law 98-99 (otherwise known as the 1985 farm bill). This act provided USDA the authority to conduct research and education in alternative agriculture or LISA (USDA, 1988).

The first report of the club of Rome titled "The limits to growth" was published in 1972 by D.H. Meadows and drew attention in particular to the problem of limited resources of raw materials.

"Our common future" was the title of Brundtland report of the world commission for environment and development that came in 1987 and pointed out main problems related to sustainability. It contributed much to preparation for the 1992 UN conference in Rio that brought problems of sustainability to its current level of attention.

Sustainability Profile

In the early 1970s, there was increasing awareness of the impact of modern (industrial) technologies on the environment as there was evidence of pesticides in food chains. Accumulation of nutrients applied to crops in stream and underground acquifers. Over exploitation of water resources depleted the water bodies. But the major set back was the energy shortage in 1970s, which reflected that earth's resources were limited. Then the thinking of alternative agriculture gained momentum with the publications of Walter's (1975). The case for eco-agriculture Rodales (1983) Breaking new ground and Jakson's (1980) New roots for agriculture, with the movement of alternative agriculture another approach stemmed up called as agroecology movement the works of Altierri (1987) and Dover and Talbot's (1987) are famous.

Mile-stones in the History (Chronological Sequence)

Vedic period, Ancient

300 BC; Arthashastra, mention of soil management, commodity trade and mention of several manures.

400 BC; Krishiparashara - the first ever text book on agriculture, specific references to soil management appeared.

1400-5500 BC: Ramayana and Mahabharata: Water management and mention of Kamadhenu, the celestial cow and its role in human life and soil fertility.

2500-1500 BC; Mention of organic manures, green manures in Vedas.

590 A.D; Holy Quran - stresses on recycling i.e. atleast one third of what is taken out from soils must be returned to it.

- In Germany <http://en.wikipedia.org/wiki/Germany>, Rudolf Steiner <http://en.wikipedia.org/wiki/Rudolf_Steiner>'s development of biodynamic agriculture <http://en.wikipedia.org/wiki/Biodynamic_agriculture> was probably the first comprehensive organic farming system (the apparent beginning of which was a lecture Steiner presented in 1924 <http://en.wikipedia.org/wiki/1924>).
- 1930's Nature farming was developed in Japan by Mokichi Okada
- In 1939 <http://en.wikipedia.org/wiki/1939>, strongly influenced by Sir Howard's work, Lady Eve Balfour <http://en.wikipedia.org/wiki/Lady_Eve_Balfour> launched the Haughley Experiment <http://en.wikipedia.org/wiki/Haughley_Experiment> on farmland in England. It was the first scientific <http://en.wikipedia.org/wiki/Scientific>, side-by-side comparison of organic and conventional farming. Four years later, she published *The Living Soil <http://en.wikipedia.org/wiki/The_Living_Soil>*, based on the initial findings of the Haughley Experiment. It was widely read, and lead to the formation of a key international organic advocacy <http://en.wikipedia.org/wiki/Advocacy> group, the Soil Association <http://en.wikipedia.org/wiki/Soil_ Association>.

- Sir Albert Howard <http://en.wikipedia.org/wiki/Sir_Albert_Howard> is often referred to as the father of modern organic agriculture. His writings, and notably, the 1940 <http://en.wikipedia.org/wiki/1940> book <http://en.wikipedia.org/wiki/Book>, *An Agricultural Testament <http://en.wikipedia.org/wiki/An_Agricultural_Testament>*, influenced many scientists and farmers of the day.
- During the 1950s <http://en.wikipedia.org/wiki/1950s>, sustainable agriculture was a research <http://en.wikipedia.org/wiki/Research> topic of interest, but science tended to concentrate on the new chemical approaches. In the U.S. <http://en.wikipedia.org/wiki/United_States>, J.I. Rodale <http://en.wikipedia.org/wiki/J.I._Rodale> began to popularize the term and methods of organic growing. In addition to agricultural research, Rodale's publications through the Rodale Press <http://en.wikipedia.org/wiki/Rodale_Press> helped to promote organic gardening <http://en.wikipedia.org/wiki/Organic_gardening> to the general public.
- In 1962 <http://en.wikipedia.org/wiki/1962>, Rachel Carson <http://en.wikipedia.org/wiki/Rachel_Carson>, a prominent scientist and naturalist <http://en.wikipedia.org/wiki/Natural_history>, published <http://en.wikipedia.org/wiki/Publish> *Silent Spring <http://en.wikipedia.org/wiki/Silent_Spring>*, chronicling the effects of DDT <http://en.wikipedia.org/wiki/DDT> and other pesticides on the environment. A bestseller <http://en.wikipedia.org/wiki/Bestseller> in many countries <http://en.wikipedia.org/wiki/Countries>, including the US, and widely read around the world, *Silent Spring* was instrumental in the US government <http://en.wikipedia.org/wiki/Government>'s banning of DDT in 1972. The book and its author are often credited with launching the environmental movement.

- 1965; Introduction of high yielding Mexican variety of wheat
- 1967; High yielding paddy varieties from IRRI (Philippines) to India.
- 1968; William Gaud of the U.S.D.A. coined the term green revolution
- In the 1970s <http://en.wikipedia.org/wiki/1970s>, worldwide movements concerned with the pollution <http://en.wikipedia.org/wiki/Pollution> and the environment increased attention on organic farming. As the distinction between organic and conventional food became clear, one goal of the organic movement was to encourage consumption of locally grown food <http://en.wikipedia.org/wiki/Local_food>, which was promoted through slogans <http://en.wikipedia.org/wiki/Slogan> such as "Know Your Farmer, Know Your Food".
- 1972; Report from the Rome club "Limits to Growth" wherein proposals for sustainable agriculture development was made.
- 1972; International Federation of organic agriculture movement was established by six organizations of three continents.
- 1978: Bill Mollison, an Australian Forest ecologist coined the term permaculture.
- 1980; U.S. Govt. announced its "Global 2000" special report wherein there is an indication of sustainable agriculture in these topics.
- 1980; The International Union of Conservation of Nature and Natural Resources (IUCN) proposed the concept of sustainability in the context of world conservation strategy.

- In the 1980s <http://en.wikipedia.org/wiki/1980s>, around the world, various farming and consumer groups began seriously pressurising for government regulation of organic production. This led to various legislation and certification standards being enacted through the 1990s <http://en.wikipedia.org/wiki/1990s> and to date. Currently, most aspects of organic food production are government-regulated in the US and the European Union.
- 1980: Based on directions from the Secretary of Agriculture (1979), the U.S. Department of Agriculture (USDA) announced the report, "Organic Agriculture in the United States - Facts and recommendations" that reported on the facts and problems with organic agriculture in the U.S., Europe and Japan.
- 1981: The FAO "Agriculture: Toward 2000" had a chapter entitled, "Sustainability in Production".
- 1984: The agricultural committee of the U.S. National Research Council (NRC) established a subcommittee for the Role of Alternative Agricultural Methods in Modern Agricultural Production, and started deliberations.
- 1985: The Technical Advisory Committee (TAC) of the CGIAR indicated that it was necessary to continuously monitor sustainability in agriculture.
- 1985: Meeting of environmental ministers, it was recognized that, of human activities, Agriculture was one of the factors that placed burden on environment.
- 1986: National Land use Policy (NLP) was outlined by National Land - Use and Wasteland Development council for India.
- 1986: Farming Act of U.S. included legislation for research and education programmes for sustainable agriculture.

- 1986: The spread of groundwater pollution caused by nitrate nitrogen throughout European countries led to considerations by the environmental policy committee concerning the relations between agriculture and environment.
- 1986-87: Launching of a centrally sponsored scheme on National watershed development programme for rainfed areas (NWDPRA) during 7th five year plan.
- 1987: The world Committee on the Environment and Development (WCED) report, Our Common Future" indicated the importance of sustainable agriculture.
- 1987: National water policy adopted by India
- 1987: The report "To protect the future of the world from the world committee on environment and development (WCED) defined the term sustainability
- 1988: The EC committee compiled a communication paper on the environment and agriculture.
- 1988: The TAC of CGIAR held the Workshop on the Sustainability of Agriculture.
- 1988 Actual research on LISA by the USDA started
- 1989: The TAC of CGIAR announced the report, "Sustainable Agricultural Development - suggestions for International agricultural Research".
- 1989: The white paper on Global Agriculture by the FAO emphasized "sustainable development and management of natural resources.
- September, 1989: The NRC announced its report "Alternative Agriculture" that suggested a change from chemical material dependent agriculture.
- September, 1989: In response to the above, the assistant undersecretary of agriculture at the USDA announced his expectations for alternative agriculture.

- October, 1989: The main subjects at the OECD agricultural research bureau chief agriculture good for the environment. There were deliberations on Low Input Sustainable Agriculture (LISA) and organic agriculture.
- 1989: Board of Agriculture of the National Research council of the U.S. published a major study called "Alternative agriculture".
- October, 1989: The official publication, "Agricultural Research of the Agricultural Research Bureau of the USDA" carried a special feature introducing recent research on sustainable agriculture.
- 1989: The conference was prompted by the report *Our Common Future <http://en.wikipedia.org/wiki/Brundtland_Report>* (1987 <http://en.wikipedia.org/wiki/1987>, World Commission on Environment and Development, also known as the Brundtland <http://en.wikipedia.org/wiki/Gro_Harlem_Brundtland> Commission), which called for strategies to strengthen efforts to promote sustainable and environmentally sound development. A series of seven UN conferences on environment and development followed. The Brundtland Commission <http://en.wikipedia.org/wiki/Brundtland_Commission> coined the most widely used definition of sustainable development, which contains two key concepts: The concept of "needs", in particular the essential needs of the world`s poor, to which overriding priority should be given; and the idea of limitations imposed by the state of technology and social organization on the environment's ability to meet the present and future needs.
- 1990: Farming Act of U.S. defined sustainable agriculture and renamed LISA to sustainable agriculture and education (SARE).

- 1991: World sustainable agricultural association was established and has activities in US, India, Japan, Thailand, Taiwan, Australia and Beijing.
- 1992: United Nationals conference on environment and development (UNCED) held in Rio (Earth summit).
- 1992: The first major manifestation of this popularization of sustainable development occurred at the United Nations Conference for Environment and Development (the Earth Summit <http://en.wikipedia.org/wiki/Earth_Summit>) in 1992 <http://en.wikipedia.org/wiki/1992>.
- 2000: a) An important year for Indian planning commission as it constituted a steering group, which identified organic farming as National challenge. b) Ministry of commerce launched the National organic programme in April.
- In the 2000s <http://en.wikipedia.org/wiki/2000s>, the market for organic products, including food, beauty, health, bodycare, and household products, and fabrics, continues to grow rapidly worldwide. More countries are establishing formal, government-regulated certification of organic food: in 2002 <http://en.wikipedia.org/wiki/2002> in the US, and projected for 2006 <http://en.wikipedia.org/wiki/2006> in Canada <http://en.wikipedia.org/wiki/Canada>, among others. Monitoring and challenging certification rules and decisions have become a regular, high profile aspect of activists in the organic movement.

❑❑❑

4 PRINCIPLES OF SUSTAINABLE AGRICULTURE

A systems perspective is essential in understanding sustainability. The system is envisioned in its broadest sense *i.e.,* from the individual farm to the local ecosystem and to communities affected by this farming system both locally and globally. An emphasis on the system allows a longer and thorough view of the consequences of farming practices on both human communities and the environment. A systems approach gives us the tools to explore the interconnections between farming and other aspects of our environment.

Towards Sustainability

Sustainability means preserving human life on Earth. Meeting human needs is thus vital in creating a sustainable society. It follows that one of The Natural Step's principles of sustainability is to *meet human needs worldwide.*

The other three principles focus on interactions between humans and the planet. They are based in science and supported by the analysis that ecosystem <http://en.wikipedia.org/wiki/Ecosystem> functions and processes are altered in the following ways:

- Society mines and disperses materials faster than they are returned to the Earth's crust <http://en.wikipedia.org/wiki/Crust_(geology)> (examples include oil, coal and metals such as mercury and lead).
- Society produces substances faster than they can be broken down by natural processes they can be broken down at all (examples of such substances include dioxins <http://en.wikipedia.org/wiki/Dioxins>, DDT <http://en.wikipedia.org/wiki/DDT> and PCBs <http://en.wikipedia.org/wiki/PCB>).
- Society depletes or degrades resources faster than they are regenerated (for example, over-harvesting of trees or fish), or by other forms of ecosystem manipulation (for example, paving over fertile <http://en.wikipedia.org/wiki/Fertile> land or causing soil erosion <http://en.wikipedia.org/wiki/Soil_erosion>).

This can be supplemented by some fundamental principles of sustainable agriculture:

- that farm productivity is enhanced over the long term;
- that adverse impacts on the natural resource base and associated ecosystems are ameliorated, minimised or avoided;
- that residues resulting from the use of chemicals in agriculture are minimised;
- that net social benefit (in both monetary and non-monetary terms) from agriculture is maximised; and
- that farming systems are sufficiently flexible to manage risks associated with the vagaries of climate and markets.

Five Basic Principles of Sustainable Agriculture

Resource conservation

Emphasize conservation of soil, water, energy and biological resources.

Use renewable sources of energy instead of non-renewable sources.

Shift from through flow nutrient management to recycling of nutrients, with increased dependence on natural processes.

Efficient and human use of inputs

Land use must be according to the natural potential of the land.

Sustainability and soil fertility

The fertility of the soil is central to the sustainability of both natural and managed ecosystems because it is the medium from which terrestrial production emanates, Greenland (1975) suggested that there are 5 basic principles of soil management that are essential for sustainable agricultural systems.

1. The chemical nutrients removed by the crops must be replenished in soil.
2. The physical condition of the soil should be maintained, which usually means that the humus level must be constant or increase.
3. There must be no soil acidity or toxic elements.
4. Soil erosion must be controlled to be equal to or less than the rate of soil genesis.
5. Management of the soil to enhance and protect soil quality.

Maintain the stability of Ecosystem

Selection of species and varieties that are well suited to the site and to conditions of the farm i.e. use multiple varieties and races of crops and animals on farm, thus avoiding dependence on single crop products. Diversification of crops (including livestock) and cultural practices to enhance the biological and economic stability of the farm, eliminate the use of non-renewable of farm human inputs that have the potential to harm the environment on the health of farmers, labourers and consumers.

When materials must be added to the system, use naturally occuring materials instead of synthetic manufactured inputs. Use of farm resources as much as possible. Manage pests, diseases and weeds instead of controlling them.

Adopt plants and animals to the ecological conditions of the farm return than modifying the farm to meet the needs of the crops and animals.

Make more appropriate matches between cropping patterns and the productive potential and physical limitations of the farm.

System / holistic principle

Value most highly the overall health of the agro-ecosystem rather than the outcome of a particular crop system or season.

There must be organic and harmonious relations among the elements that makeup the system.

The inter-relatedness of all parts of a farming system, including the farmer and his family.

Ensure food safety / quality

By eliminating environmental pollution by toxic and surplus nutrients and ensure human, cultural and environmental health. This produce foods of high nutritional quality and sufficient quality.

Visualize sustainability

Incorporate the idea of long term sustainability into overall agroecosystem design and management.

Use long term strategies and develop plans that can be adjusted and reevaluated through time & build soil organic matter that will ensure soil fertility build up over the long term.

The farm productivity is enhanced over the long term.

Provisions should be made to safeguard resources for future generations.

❑❑❑

5 RESOURCE CONSERVATION AND ENVIRONMENTAL ISSUES

Sustainable agriculture can be defined as a tool of using the natural resources for meeting the present needs without jeopardizing the future potential, increasing per capita productivity without causing degradation of soil and water resources, enhancing profitability without reducing total production and reducing excessive dependence on off farm inputs. Thus the holistic view of sustainability is an approach which is ecologically sound, economically profitable and socially equitable. Theoretically this is an ideal concept and of recent origin. But the real danger to sustainable agriculture stem from the compelling needs of increasing population for food, fuel, fodder on one hand and limited natural resource base of land and water on the other.

Characterisation of Natural Resources

The objective of food security can be met by promoting sustainable agricultural production systems. Increasing agricultural production should not be at the cost of degrading soil and depicting the water bodies. If the basic needs of people are to be met on a sustainable basis, the earth's natural resource base must be conserved and enhanced. In other words, for food security and sustainable livelihoods, ecological security is necessary. Areas under ecological stress need attention, pressure on natural resources increases with population increase in order to fulfill the demand.

In environmental terms, for a system to be sustainable, it must exhibit two characteristics.

1. It must not deplete natural resources.
2. It must not create an environmental pollution overload.

Sustainable agricultural systems are designed to take maximum advantage of existing soil nutrient and water cycles, energy flows and soil organisms for food production. As well, such systems aim to produce food that is nutritious, without being contaminated with products that might harm human health.

Land

Among the major resources available to mankind is land comprising of soil, water, associated plants and animals. The terms soil and land are not synonymous but inter-related. Land is the term connoting all the features of man's national geographical environment with which he deals under economic motivation to raise a variety of crops. The basic input for crop production is the land, which is a vital natural resource. The physical losses of such land as a result of land management are on the increase. Planet earth of 15.2 billion ha. give 3.8 ha per person.

In land, the top soil carries the maximum value. A layer of 20 cm deep top soil produces 99% of food required for the living organism in the world. To keep this skin of the earth in good health and under constant improvement is basic to all life.

Available land resources

At global level, around 5 billion people are currently being sustained on 13.4 billion ha geographical area, of which only around 1.4 billion ha are under cultivation. India is sustaining 16 per cent of global population on 2.4 per cent of world's geographical area.

The geographical area of India is 328.73 m ha, consisting mainly of red sandy, medium black, alluvial and loamy soil.

Out of 328.73 m ha, the reporting area for land utilization statistics is 304.34 m ha and details of the land use classification are mentioned in Table. From the table, it can be observed that our cultivated area has remained constant at 140 m ha since last three decades. Further, it is observed that of the cultivated area only 42 m ha is irrigated (30%) and the balance 70 per cent area in the country is still cultivated under rainfed conditions. As many as 128 districts in the country fall under the category of dry farming where the monsoon is erratic and they produce most of the coarse cereals, pulses, oilseeds, cotton and dry fodder.

Water

Water is essential for human beings, animals and plants. Without water life is impossible on earth *i.e.,* water and life are inseparable. The global total water resources are approximated to be about 1,384 million km^3. About 97.16 per cent of total water in the world occurs in the oceans as salt water. Only 2.84 per cent is fresh water. According to a rough estimate the annual global flow of water ranges between 4000 to 4700 m ha m of which 1400 m ha m is utilisable for irrigation and other purposes.

India is a vast country and receives an annual average rainfall of 1190 mm. The rainfall variation is very high (in the country) which is from 10,000 mm in Khasi and Jaintia hill areas and only 100 - 150 mm in western Rajasthan.

Unequal geographical distribution, unequal distribution and frequent departures from the normal rainfall characterize, the rainfall of this country. The pattern of rainfall indicates that this important source of water is not available uniformly over different parts of the country. In some parts, rain water is received in excess quantity and a major part of it is lost by runoff, while in low rainfall areas it is not sufficient to meet the daily evaporative demand of the atmosphere. Thus under uncertainity of rainfall, development of irrigation potential and water conservation assumes importance.

The average annual precipitation over the whole country is 1190 mm, when geographical area of 328 m ha (India) receives 1190 mm, this rainfall in terms of volume amounts to 392 m ha m, which may be rounded of to 400 m ha m including snowfall. Irrigation commission of India (1972) has placed the total annual surface flow in the country at 180 m ha m. On full harnessing and mobilization of these resources by 2025 AD, it is envisaged that 70 m ha m of surface water and 35 m ha m of ground water can be mobilised for irrigation. The projected use of this 105 m ha m of water is 77 m ha m for irrigation and 28 m ha m for domestic and industrial water supply and all other purposes. The major use of water in the country is for irrigation (85%). The demand for water is increasing with expanding population, livestock, industries, etc. It is estimated that 12 m ha m of water is needed for industry by 2025 AD compared to 1.0 m ha m used in 1985. The cost of providing irrigation facilities is very high. At present the cost of providing irrigation for one ha of land under major surface irrigation is over Rs. 43,000. The available water resources have to be very efficiently utilised by minimising water losses in the system.

Biodiversity

Agro biodiversity is a fundamental feature of farming systems around the world. It encompasses many types of biological resources tied to agriculture, including;

- genetic resources - the essential living materials of plants and animals;
- edible plants and crops, including traditional varieties, cultivars, hybrids, and other genetic material developed by breeders; and
- livestock (small and large, lineal breeds or thorough breds) and fresh water fish;
- soil organisms vital to soil fertility, structure, quality and soil health;
- naturally occurring insects, bacteria and fungi that control insect pests and diseases of domesticated plants and animals;
- Agro ecosystem components and types (polycultural/ monocultural, small/ large scale, rainfed/ irrigated, etc.) indispensable for nutrient cycling, stability and productivity and
- Wild resources (species and elements) of natural habitats and landscapes that can provide services (for example, pest control and ecosystem stability) to agriculture.

Biodiversity is now recognised worldwide as a major factor in identification of novel gene pools. Experts have described only about 1.5 million species (Table) by biosystematics so far and they have calculated that another 70-80 million species exist if we study invertebrates and micro-organisms more exhaustively. Many of these natural populations are either endangered or becoming extinct (Table) owing to natural and man-made environmental changes. Conservation of the heritage in well established germplasm banks is vital (Table) to further development in agriculture.

Table: Numbers of described species of living organisms

Organism	Number
Viruses	1,000
Bacteria + blue - green algae	4,760
Fungi	46,983
Algae	26,900
Lower plants	28,426
Higher plants	2,20,000
Protozoans	30,800
Fish	19,056
Birds	1,940
Mammals	4,000
Total	**13,92,485**

Table: Plant species spectrum

Nature of species	Number of species
Existence	240,000
Endangering species	20,000 – 25,000
Analyzed species	5,000
Cultivable species	3,000
Caloric species	30
Consumption species (wheat, corn, rice & potato)	4

Source: T.N. Khoshoo (1992)

Table: Biodiversity of germplasm collection in agriculture

Rice	40,000
Wheat	40,650
Pulses	41,350
Oilseeds	30,415
Cotton	20,755
Coffee	460
Tea	1,725
Vegetables (summer crop)	3,650
Vegetables (winter crop)	1,850
Total	**2,16,075**

Government of India helped to establish a germplasm collection center for field crops and named it as National Bureau of Plant Genetic Resources (NBPGR). The accessions are pervasive and broad spectrum and hence serve the purpose of breeders and researchers of plant breeding and genetics (Table). The *in situ* and *ex situ* conservation programmes are initiated in India with the identification of biosphere reserves (Table).

Table: *In situ* conservation in India

Region	Biosphere reserve	Area (km2)	State involved
Himalayas	Nandhadevi	1,560	Uttar Pradesh
Myanmar monsoon forests	Nokrek	80	Meghalaya
Bengalian	Manas	2,837	Assam
Rainforest	Sundarbans	9,630	West Bengal
Coramandal	Gulf of Mannar	555	Tamil Nadu
Malabar	Nilgiris	5,620	Karnataka, Kerala and Tamil Nadu
Andaman/ Nicobar	Great Nicobar	885	Andaman & Nicobar islands

Table: *Ex situ* Conservation in India by NBPGR

Centre	Strategy
Field Gene Banks	Field level regeneration
National Gene Banks species	Long time storage for orthodox at -20° C
National Plant Tissue Culture Repository	Tissue culture concept for recalcitrant species
National Herbarium of Cultivated Plants and Wild Relatives	Authentication of specimens

Biodiversity benefits

Utilitarian value

Undoubtedly, there is a tremendous, undiscovered wealth of biological products that are of potential use to humans. Many of these natural products are present in the biodiversity of tropical species that have not yet been "discovered" by taxonomists. There are utilitarian reasons and we must take advantage of biodiversity in myriad ways for sustenance, medicine, shelter and other purposes. If species become extinct, their unique services, be they biological, ecological or otherwise are no longer available for exploitation. There are many cases where research on previously unexploited species of plants and animals has revealed the existence of products of great utility to humans, such as food or medicinals.

Diversity and ecosystem stability

The major benefit of biodiversity is the stability and integrity of ecosystems, *i.e.*, in terms of preventing erosion and controlling nutrient cycling, productivity, tropic dynamics and other aspects of ecosystem structure and function.

Each species of living organism occupies a particular habitat and serves a particular function in an ecosystem. The function and habitat constitute the organism's ecological niche. A basic characteristic of a healthy or well-balanced ecosystem is an overlapping of niches occupied by different species. A stable ecosystem can withstand some external stress, such as pollution, construction, or hunting without being completely disrupted or damaged.

In a stable ecosystem, if any one species disappears because of natural or artificial causes, other species are available to occupy its niche and take over its role in the food chain. Actually, the term food web is more appropriate for a healthy ecosystem because of the overlapping nature and complexity of the eat-and-be-eaten by relationships.

A tropical rain forest is a good example of a stable ecosystem because of the tremendous number of plant and animal species thriving in it. The loss of one species of tree or one species of animal is not likely to have a significant impact on the whole ecosystem.

Interconnectedness of the living world

The rich diversity of the living world is connected in two distinct ways. First, different types of organisms live side by side in complex ecological networks of interdependency, each relying on the others that share its habitat for nutrients and energy. Second, all life on Earth is connected in an evolutionary tree of life. At the bottom of the tree is the common ancestor from which all living things descended a single-celled microbe that lived more than 3.5 billion years ago and in its uppermost branches are gorillas, chimpanzees, orangutans, and our own species, *Homo sapiens*.

Ecological diversity

Ecological diversity is the intricate network of different species present in local ecosystems and the dynamic interplay between them. An ecosystem consists of organisms from many different species living together in a region that are connected by the flow of energy, nutrients, and matter that occurs as the organisms of different species interact with one another. The ultimate source of energy in nearly all ecosystems is the Sun. The Sun's radiant energy is converted to chemical energy by plants. This energy flows through the systems when animals eat the plants and then are eaten, in turn, by other animals. Fungi derive energy by decomposing organics, releasing nutrients back into the soil as they do so. An ecosystem, then, is a collection of living components microbes, plants, animals, and fungi and nonliving components climate and chemicals that are connected by energy flow.

Evolutionary diversity

Every species on Earth is related to every other species in a pattern every bit as complex as the patterns of energy flow within an ecosystem. In evolutionary diversity, the connection is not energy flow, but rather genetic connections that unite species. The more closely related any two species are, the more genetic information they will share, and the more similar they will appear. An ever-widening circle of evolutionary relatedness embraces every species on Earth.

This is the evolutionary chain of life. All species are descended from a single common ancestor. From that ancient single-celled microbe, all inherited RNA. As time goes by, species diverge and develop their own peculiar attributes, thus making their own contribution to biodiversity.

Global biodiversity crisis

Most biologists accept the estimate of American evolutionary biologist Edward O. Wilson that the Earth is losing approximately 27,000 species per year. This estimate is based primarily on the rate of disappearance of ecosystems, especially tropical forests and grasslands, and our knowledge of the species that live in such systems. We can measure the rate of loss of tropical rain forests, for example, by analyzing satellite photographs of continents from different periods that show rates and amounts of habitat destruction and from these measurements calculate the approximate number of species being lost each year.

This extraordinary rate of extinction has occurred only five times before in the history of complex life on Earth. Mass extinctions of the geological past were caused by catastrophic physical disasters, such as climate changes or meteorite impacts, which destroyed and disrupted ecosystems around the globe. In the fifth mass extinction, which occurred more than 65 million years ago, the Earth was shrouded in a cloud of atmospheric dust as a result of meteorite impact or widespread volcanic activity. The resulting environmental disruption caused the demise of 76 percent of all species alive at the time, including the dinosaurs. Today's sixth extinction is likewise primarily caused by ecosystem disturbance but this time the destroying force is not the physical environment, but rather humankind. The human transformation of the Earth's surface threatens to be every bit as destructive as any of the past cataclysmic physical disasters.

Human Impact

The underlying cause of biodiversity loss is the explosion in human population, now at 6 billion, but expected to double again by the year 2050. The human population already consumes nearly half of all the food, crops, medicines, and other useful items produced by the Earth's organisms, and more than 1 billion people on Earth lack adequate supplies of fresh water. But the problem is not sheer numbers of people alone: The unequal distribution and consumption of resources and other forms of wealth on the planet must also be considered. According to some estimates, the average middle-class American consumes an amazing 30 times what a person living in a developing nation consumes. Thus the impact of the 270 million American people must be multiplied by 30 to derive an accurate comparative estimate of the industrialized nations have on the world's ecosystems.

The single greatest threat to global biodiversity is the human destruction of natural habitats. Since the invention of agriculture about 10,000 years ago, the human population has increased from approximately 5 million to a full 6 billion people. During that time, but especially in the past several centuries, humans have radically transformed the face of planet Earth. The conversion of forests, grasslands and wetlands for agricultural purposes, coupled with the multiplication and growth of urban centers and the building of dams and canals, highways, and railways, has physically altered ecosystems to the point that extinction of species has reached its current alarming pace.

In addition, overexploitation of the world's natural resources, such as fisheries and forests, has greatly outstripped the rate at which these systems can recover. For example, 12 of the 13 largest oceanic fisheries are severely depleted. Modern fishing techniques, such as using huge fishing nets and bottom vacuuming techniques, remove everything in their paths including tons of fish and invertebrates of no commercial use.

As human populations have grown, people have spread out to the four corners of the Earth. In the process, whether on purpose or by accident, they have introduced non-native species that have created ecological nightmares, disrupting local ecosystems and, in many cases, directly driving native species extinct. For example, the brown tree snake was introduced to the island of Guam, probably as a stow away on visiting military cargo ships after World War II (1939-1945). The snake devastated the native bird population, driving over half a dozen native species of birds to extinction simply because the native birds had not been exposed to this type of predator and did not recognize the danger posed by these snakes.

Preserving biodiversity

The biodiversity crisis is a very real and very important aspect of the global environmental crisis. All nations have a responsibility to maintain biodiversity within their own jurisdictions and to aid nations with less economic and scientific capability to maintain their biodiversity on behalf of the entire planet. The modern biodiversity crisis focuses on species rich tropical ecosystems, but the developed nations of temperate latitudes also have a large stake in the outcome and will have to substantially subsidize global conservation activities if these are to be successful. Much needs to be done, but an encouraging level of activity in the conservation and protection or biodiversity is beginning in many countries, including an emerging commitment by many nations to the conservation of threatened ecosystems in the tropics.

As the scope and significance of biodiversity loss become better understood, positive steps to stem the tide of the sixth extinction have been proposed and to some extent, adopted. Several nations have enacted laws protecting endangered wildlife. An international treaty known as the Conservation on the International Trade of Endangered Species (CITES) went into effect in 1975 to outlaw the trade of endangered animals and animal parts. The Convention on Biological Diversity, held in Rio de Janeiro, Brazil, in 1992 and ratified by more than 160 countries, obligates governments to take action to protect plant and animal species conservation biologists also work with established industries to develop practices that ensure the health and the sustainability of many fish the fishers can harvest without damaging the population and the ecosystem as a whole and same principles are applied to the harvesting of trees, plants, animals and other natural resources.

In the last three decades, focus has shifted away from the preservation of individual species to the protection of large tracts of habitats linked by corridors that enable animals to move between the habitats. Thus the movement to save, for example, the spotted owl of the Pacific Northwest, has become an effort to protect vast tracts of old-growth timber.

Promising as these approaches may be, conservation efforts will never succeed in the long run if the local economic needs of people living in and near threatened ecosystems are not taken into account. This is particularly true in developing countries, where much of the world's remaining undisturbed land is located. At the end of the 20th century, international organizations such as the World Bank and the World Wildlife Fund launched a movement for all countries in the developing world to set aside 10 per cent of their forests in protected areas. But many communities living near these protected areas have relied on the rain forest for food and firewood for thousands of years. Left with few economic alternatives, these communities may be left without enough food to eat.

Preserving biodiversity also takes place at the molecular level in the conservation of genetic diversity. All around the world efforts are being made to collect and preserve endangered organisms' DNA, the molecule that contains their genes. These collections, or *gene banks*, may consist of frozen samples of blood or tissue, or in some cases, they may consist of live organisms. Biologists use gene banks to broaden the gene pool of a species, increasing the likelihood that it will adapt to meet the environmental challenges that confront it. Many zoos, aquariums and botanical gardens work together to carefully maintain the genetic diversity in captive populations of endangered animals and plants, such as the giant panda, the orangutan, or the rosy periwinkle. Captive animals are bred with wild populations, or occasionally released in hopes that they will breed freely with members of the wild population, thus increasing its genetic diversity. These gene banks are also an essential resource to replenish the genetic diversity of crops, enabling plant breeders and bioengineers to strengthen their stocks against disease and changing climatic conditions.

Biodiversity underlies everything from food production to medical research. Humans, the world over use of only 40,000 species of plants and animals on a daily basis. Many people around the world still depend on wild species for some or all of their food, shelter and clothing. All of our domesticated plants and animals came from wild-living ancestral species (*Courtsey;* Microsoft, Encarta 205).

Energy

Energy is the lifeblood of ecosystems and of the biosphere as a whole. At the most fundamental level, what ecosystems do is capture and transform energy.

Energy is constantly flowing through ecosystems in one direction. It enters as solar energy and is converted by photosynthesizing organisms (plants and algae) into potential energy, which is stored in the chemical bonds of organic molecules, or biomass. Whenever this potential energy is harvested by organisms to do work (*e.g.* grow, move, reproduce), much of it is transformed into heat energy that is no longer available for further work or transformation - it is lost from the ecosystem.

Agriculture, in essence, is the human manipulation of the capture and flow of energy in ecosystems. Humans use agroecosystems to convert solar energy into particular forms of biomass - forms that can be used as food, feed, fibre and fuel.

All agroecosystems - from the simple, localized plantings and harvests of the earliest agriculture to the intensively altered agroecosystems of today - require an input of energy from their human stewards in addition to that provided by the sun. This input is necessary in part because of the heavy removal of energy from agroecosystems in the form of harvested material. But it is also necessary because an agroecosystem must to some extent deviate from and be in opposition to, natural processes.

The agricultural "modernization" of the last several decades has been largely a process of putting ever greater amounts of energy into agriculture in order to increase yields. But most of this additional energy input comes directly or indirectly from non-renewable fossil fuels.

Moreover, the return on the energy investment, in conventional agriculture is not very favourable; for many crops, we invest more energy than we get back as food. Our energy intensive form of agriculture, therefore, cannot be sustained into the future without fundamental changes.

Energy is vital for economic and social development and comes from fossil fuel reserves. In India, fossil fuel reserves are estimated (Table) and it seems that the oil source will last for next 20 years. But the line of energy demand goes high vide the increase of usage. The energy consumption pattern of India though comparatively low in the world scenario (Table), the corporate sector consumption is equal to their counterparts of developed countries (Table). To meet the increasing demand for energy in the rural India very limited options are available. Since Indian farmers are wedded to their cattle muscle power, available dung finds its way to chulah. Having realized the fact that both energy and manures can be derived from such dung and associate safe solid waste disposal. National Project for Biogas Development (NPBD) is implemented. Methane rich biogas is a safe cooking gas. India has been one of the pioneering countries in the development and use of biogas technology.

Table. Fossil Fuel Reserves

Coal	180 billion tonnes
Lignite	5,900 million tonnes
Oil	638.44 million tonnes
Gas	579.47 billion m^3

Table. Energy Consumption Pattern

Region	Per capita consumption
India	272
Africa	424
Asia	689
South America	1,012
North America	6,667
Europe	4,440
World	1,896

Optimal Use and Management of Energy

Energy plays an important role in promoting and sustaining development. Research studies conducted in India and other countries have shown that energy use per hectare is positively correlated with crop yields per hectare. The same conclusion holds in the case of industrial production. Availability of a safe and sustainable source or a mix of source of energy is a critical determinant of sustainable development. The key elements of sustainability that have to be reconciled are (a) sufficient growth of energy supplies to meet the growing needs of Indian agriculture and industry; (b) energy efficiency and conservation measures such that the waste of primary resources is minimised; and (c) protection of the biosphere and prevention of more localised forms of pollution (WCED, 1987).

The world's consumption of commercial energy from conventional sources is increasing at a rapid rate depicting the world's reserve of fossil fuels. As per the projections made by the International Energy Agency, by 2010, global energy consumption is expected to rise by about 50 per cent from the 1993 level (WRI, 1998). It is also estimated that the world is now left with only 90 years of proved recoverable mineral reserves 243 years of proved reserves in place and 800 years of total reserves (UN, 1992). Besides, there are wide disparities in energy consumption between regions and countries. It is estimated that in 1989 the per capita energy consumption in industralised countries was 10 times the average per capita consumption in the developing countries (WRI, 1992).

The final energy consumption in India has grown at the rate of 5.5 per cent per annum. But even now, the per capita consumption of commercial energy is very low when compared with that in developed countries. For example, in 1996, it was 248 kg of oil equivalent which was only 3 per cent of the per capita consumption in U.S.A. and about 9 per cent of that in Korea (TERI, 1997). Agricultural uses accounted for about 31 per cent of the total energy consumption in the country in 1994-95 (Government of India, 1997). But given its very large population that is rapidly increasing and growing industrialisation and commercialisation of agriculture, even this low level of energy consumption is not sustainable unless non-conventional renewable sources of energy are fully developed and harnessed. It is estimated that at the current level of production, India's coal reserves will last about 45 years, oil by 21 years, and natural gas by 38 years. Besides, India has assured uranium resources of 34,000 m tonnes of which 5,000 m tonnes are economically exploitable. It also has a huge potential for producing hydro-electricity estimated at 84,000 MW of which only 24 per cent has been exploited so far. But on the whole, India is not self-sufficient in meeting its energy requirements and is a net importer of energy. In 1994-95, it imported 41.7 m mt of crude oil and petroleum products (TERI, 1997). Now there is a lot of emphasis being placed by the government on the development of renewable sources of energy such as biogas, wind power and solar radiation. It seems that the energy security is going to be as important for India in future as food security.

Capture of Solar Energy

The starting point in the flow of energy through ecosystems and agroecosystems is the sun. The energy emitted by the sun is captured by plants and converted to stored chemical energy through the photosynthetic process. The energy accumulated by plants through photosynthesis is called primary production because it is the first and most basic form of energy storage in an ecosystem. Energy left after the respiration needed to maintain plants is net primary production (NPP) and remains as stored biomass.

Through agriculture, we can concentrate this stored energy in biomass that can be harvested and utilized, either by consuming it directly or by feeding it to animals that can give different by products for human needs or use to do work for the benefit of man kind.

Plants vary in how efficiently they can capture solar energy and convert it to stored biomass. This variation is the result of differences in plant morphology (*e.g.* leaf area), photosynthetic efficiency and physiology. It also depends on the conditions under which the plant is grown. Agricultural plants are some of the most efficient plants, but even in their case the efficient plants, but even in their case the efficiency of their conversion of sunlight to biomass rarely exceeds 1% (1% efficiency means that 1% of the solar energy reaching the plant is converted to biomass).

Although all the energy in the food we consume comes originally from the sun, additional energy is needed to produce the food in the context of an agroecosystem. This additional energy comes in the form of human labour, animal labour and the work done by machines. Energy is also required to produce the machines, tools, seed, and fertilizer to provide irrigation, to process the food and to transport it to market. We must examine all these energy inputs to understand the energy costs of agriculture and to develop a basis for more sustainable use of energy in agriculture.

It is helpful, first of all, to distinguish between the different types of energy inputs in agriculture. The primary distinction is between energy inputs from solar radiation, called ecological energy inputs and those derived from human sources, called as cultural energy inputs. Cultural energy inputs can be further divided into biological inputs and industrial inputs. Biological inputs come directly from organisms and include human labour, animal labour and manure, industrial inputs of energy are derived from fossil fuels, radioactive fission, and geothermal and hydrological sources.

It is important to note that even though we are referring to all these sources of energy as "inputs", cultural energy of either form can be derived from sources within a particular agroecosystem, such "internal inputs" of energy include the labour of farm residents, the manure of on-farm animals, and energy from on-farm windmills or wind driven turbines.

Future Energy Conservation

Clearly, sustainable food production depends to a large extent on more efficient use of energy, as well as less reliance on industrial cultural energy inputs and fossil fuels in particular. The key to more sustainable use of energy in agriculture lies in expanding the use of biological cultural energy. Biological inputs are not only renewable, they have the advantages of being locally available and locally controlled environmentally and able to contribute to the ecological soundness of agroecosystems.

Strategies for Energy Conservation

Reduce the use of industrial cultural energy, especially nonrenewable or contaminating sources such as fossil fuels

- Use minimum or reduced tillage systems that require less mechanized cultivation.
- Employ practices that reduce water use and water loss in order to reduce the amount of energy expended for irrigation.
- Use appropriate crop rotations and associations that stimulate recovery from the disturbance caused by each cropping cycle without the need for artificial inputs.

- Develop renewable, energy-efficient industrial cultural sources and uses of energy to replace fossil fuels.
- Develop on-farm sources of industrial cultural energy (*e.g* electricity, wind energy, small scale hydropower) wherever possible.
- Use industrial cultural energy more efficiently by reducing waste and making more appropriate matches between the energy's quality and its use.
- Reduce energy use in the agricultural sector by regionalizing production, and putting consumers and producers more directly in contact both seasonally and geographically.

Increase the use of biological cultural energy

- View human energy as an integral part of energy flow in agriculture rather than as an economic cost that must be reduced or eliminated.
- Return harvested nutrients to the farm land from which they came.
- Make more extensive use of manures to maintain soil fertility and quality.
- Increase the local and on-farm use of agricultural products in order to lessen the energy costs of long distance transport.
- Expand the use of biological control and integrated pest management.
- Encourage the presence of mycorrhizal relationships in the roots of crops in order to lessen the need for external inputs.

Model agroecosystems which require lower levels of cultural energy inputs

- Make greater use of nitrogen fixing crops, green manures and fallows.

- Make greater use of biological pest management through cover-cropping, intercropping, encouragement of beneficials, etc.
- Introduce crops that are appropriate or adapted to the local environment rather than trying to alter the environment to meet the needs of the crop.
- Incorporate wind breaks, hedge rows and non crop areas into cropping systems for habitat and microclimate management.
- Design agroecosystems using local natural ecosystems as a model.
- Maximize the use of successional development in the cropping system (*e.g.*, through agroforestry) in order to maintain better agroecosystem regeneration capacity (*Courtesy:* Gliessmann, 1997).

Impact of Green Revolution on Environment

Food production scenario

The changes in food production are presented in Table. For the food grains as a whole, the per cent increase in area is 0.67 per cent while the corresponding increase in production and productivity is 35.99 per cent and 35.09 per cent, respectively.

Amongst the food grains, rice and wheat have shown maximum increase. Out of the total food grain production, increase of 46.64 million tonnes during 1980-81, the contribution of rice and wheat together is 39.17 million tonnes (80.4 per cent of total increase).

But oilseeds front has witnessed a tremendous jump from 1980-81 to 1990-91 where in 97 per cent increase was observed. It is because of the foresight of the then Prime Minister late Sri. Rajiv Gandhi who launched a technology mission on oilseeds in 1985-86 and this concerted effort has yielded positive results and reduced the import bill in 1990-91.

Table. Increase in Area, Production and Productivity of crops during 1980-81 to 2003-04.

Crop	Area (million ha)				Production (mt)				Productivity (kg/ha)			
	1980-81	1990-91	2003-04	Per cent change	1980-81	1990-91	2003-04	Per cent change	1980-81	1990-91	2003-04	Per cent change
Rice	40.2	42.6	40.2		53.6	74.6	87.1		1,336	1,751	2960	
Wheat	22.3	23.9	24.9		36.3	54.5	74.5		1,630	2,274	2740	
Sorghum	15.8	14.5	9.9		10.2	11.9	7.66		660	819	772	
Millet	11.7	10.5	10.2		5.3	6.9	7.2		458	661	582	
Maize	6.0	5.9	6.3		6.9	9.1	14.87		1,159	1,524	1800	
Pulses	22.5	24.4	24.45		10.6	14.1	15.01		473	576	623	
Oilseeds	17.6	24.0	21.22		9.4	18.5	15.06		532	769	710	
Cotton	8.1	7.4	8.96		7.4	9.8	10.8		152	224	440	
Sugarcane	2.7	3.7	4.36		154.3	240.3	281.57		57,844	65,269	64562	
Potato	0.7	0.9	----		9.6	15.2	----		13,258	16,159	----	
Onion	0.25	0.3	----		2.5	3.5	----		9,961	10,605	----	
Food grains	126.7	127.5	122.5		129.6	176.2	183.17		1,023	1,382	1723	

The credit for all round development in agriculture sector should be attributed to the efforts of planners, policy makers, scientists and untiring efforts of our farmers.

Environmental degradation

Environmental degradation refers to the diminishment of a local ecosystem or the biosphere as a whole due to human activity. Environmental degradation occurs when nature's resources *viz.*, trees, habitat, earth, water, air are being consumed faster than nature can replenish them. An unsustainable situation occurs when natural capital (the sum total of nature's resources), is used up faster than that it can be replenished. Sustainability requires human activity, at its minimum, and uses only the nature's resources to the point where they can be replenished naturally:

- When it results in human consumption of renewable resources > Nature's ability to replenish than environmental degradation
- When it results in human consumption of renewable resources = Nature's ability to replenish than environmental equilibrium / sustainable growth.
- When it results in human consumption of renewable resources < Nature's ability to replenish than environmental renewal / also sustainable growth.

The long term final result of environmental degradation will be local environments that are no longer able to sustain human populations.

The recent intensification of agriculture, and the prospects of future intensification, will have major detrimental impacts on the nonagricultural land and aquatic ecosystems of the world. The first doubling of agricultural food production was associated with a 6.87-fold increase in nitrogen fertilization, a 3.48-fold increase in phosphorus fertilization, a 1.68-fold increase in the amount of irrigated cropland, and a 1.1-fold increase in land in cultivation. Based on a simple linear extension of past trends, the anticipated next doubling of global food production would be associated with approximately 3-fold increases in nitrogen and phosphorus fertilization rates, a doubling of the irrigated land area, and an increase of 18% in cropland. These projected changes would have dramatic impacts on the diversity, composition, and functioning of the remaining natural ecosystems of the world, and on their ability to provide society with a variety of essential ecosystem services. The largest impacts would be on freshwater and marine ecosystems, which would be greatly eutrophied by high rates of nitrogen and phosphorus release from agricultural fields. Aquatic nutrient eutrophication can lead to loss of biodiversity, outbreak of nuisance species, shifts in the structure of food chains and impairment of fisheries. Because of aerial redistribution of various forms of nitrogen, agricultural intensification also would eutrophy many natural terrestrial ecosystems and contribute to atmospheric accumulation of greenhouse gases. These detrimental environmental impacts of agriculture can be minimized only if there is much more efficient use and recycling of nitrogen and phosphorus in agroecosystems.

The Green Revolution is an effort to increase the food production stemming from the improved genetic strains of wheat, rice, maize and other cereals in the 1960s developed by Dr Norman Borlaug and others under the sponsorship of the Rockefeller Foundation and other organizations. This caused increase in the crop yield in India, Pakistan, Philippines, Mexico, Sri Lanka and other underdeveloped countries, preventing large scale famine.

The Green Revolution in India was initiated primarily to boost agricultural production with the purpose of feeding the burgeoning population of the country. It did succeed in doing so, but it brought with it many environmental and ecological problems. There was a degradation of land resources, decline in bio-diversity, population of water bodies and an impact on the health of human beings and other fauna and flora.

The Green Revolution is ranked as one of the greatest scientific and social achievements of the century (Task-force on Global Research on Environment and Agriculture News [GREAN, 1995]). The introduction and rapid spread of high-yielding rice and wheat varieties in the late 1960s and early 1970s resulted in a phenomenal growth in grain output. The high yields from the semi-dwarf breeding lines of rice produced by International Rice Research Institute (IRRI) and for wheat produced by Centro International de Mejoramiento de Maiz y Trigo (CIMMYT) triggered the Green Revolution (IRRI, 1996).

Aggregate rice output in Asia rose to 2.9 per cent/ annum in the period 1965-80 and from 2.1 per cent/ annum during 1955-65. Likewise, the area of rice increased.

History

William Gaud of the U.S. Department of Agriculture coined the term green revolution in March 1968 to symbolize the quantum jump in wheat production and productivity achieved in India and Pakistan during that year and to mark the beginning of the advance in rice production triggered by the semi-dwarf, non lodging varieties of rice stored in Taiwan and at IRRI in Philippines.

The revolution began in 1945 when the Rockefeller Foundation and the Mexican government established the Cooperative Wheat Research and Production Programme to improve the agricultural output of the country's farms. Norman Borlaug was instrumental in this program. This programme was continued in India and Pakistan where it is credited with saving milions of people from starvation.

The success in increasing yields was undisputable. The growth of crop yields was such that the agriculture was now able to outstrip population growth as per capita production increased every year since 1950. This growth in production due to high yielding varieties of staples such as wheat and rice has, however, drop in yields from other indigenous crops, including pulses.

Technologies

The Green Revolution technologies broadly fall into two major categories. The first is the breeding of new plant varieties; the second is the application of modern agricultural techniques in new areas.

Hybrid strains

Most crops consumed by the public-at-large in industrialized nations are Green Revolution crops. The design of high yielding varieties or hybrid strains was motivated by a desire to increase crop yield, and also to increase durability transport and longevity for storage. Norin 10 wheat is an example of such a strain that helped developing countries such as India and Pakistan to increase the productivity of their crops. Since then, strains have been bred for better appearance.

Since improved crop yield was produced mostly through the use of heavy fossil fuel inputs the increased efficiency of hybrid strains is geared towards these inputs; that is, the strains are more responding to using chemical fertilizers and also they are designed for easy mechanised harvesting.

Agricultural techniques

The techniques introduced to the developing world by the Green Revolution are :

- *Extensive use of chemical fertilizers:* Every plant basically relies on several elements to grow normally. Among them nitrogen is primary need. Only in the nitrate form can plants absorb the nitrogen they require, with the exception of rice, which can absorb ammonical nitrogen as well. Certain microorganisms found in the soil are able to convert atmospheric nitrogen into the nitrate form which plant can use for their growth. Also, some biological nitrogen fixation can take place by microorganisms living in small nodules on the roots of certain plants, such as legumes. Phosphates are also important, as well as numerous trace elements. Soil pH (acidity or alkalinity) must also be adjusted to the optimal conditions for the crop in question. Previously proper soil conditions had relied only on techniques such as crop rotation, mixing of crops, or organic manures.

- *Irrigation :* Although it has been in use in agriculture for thousands of years, the Green Revolution further developed irrigation methods to allow for more efficient irrigation. It was possible to have more than one harvest per year with reduced dependence on monsoon seasons.
- *Use of heavy machinery :* Mechanized harvesters and other machinery were not new to agriculture but the Green Revolution allowed a drastic reduction in the input of human labor to agriculture by extending the use of machinery to automate every possible agricultural process.
- *Pesticides and herbicides :* The development of chemical pesticides and herbicides (including organochlorine and organophosphate compounds) allowed further improvements in crop yields by allowing for efficient weed control (by use of herbicide early in the growing season) and eradication of insect pests.

Achievements of the green revolution

Increased yields

Green Revolution techniques have increased the production per unit area of wheat and other food crops in some major developing countries like India. Because of this, food security of many developing countries of the world has increased.

The Green Revolution in India resulted in a record grain output of 131 million tonnes in 1978-79 which made India as one of the world's biggest agricultural producer. No other country in the world which attempted the Green Revolution recorded such level of success. India also became an exporter of food grains.

The Green Revolution in agriculture helped food production to keep pace with population growth. Without the Green Revolution, agriculture would not have been able to meet the basic food requirements of the world's current population. It is estimated that the Green Revolution has saved almost a billion human lives.

Labour saving

The high level of mechanisation associated with Green Revolution techniques led to a reduced dependence on low-skilled human labour. As a result, the incomes of farmer and agricultural worker rose substantially and production costs plummeted. The efflux of labour, however, brought problems at its own, like the increased migration to the cities and creation of massive slums.

Overall, the adoption of farmers of miracle rice and wheat varieties increased food security and rural employment while lowering prices. The Green Revolution brought the following benefits to developing countries.

- Increased income and nutritional status of the poor by reducing food prices:

- Enhanced productivity.
- Freed labour for non-agriculture sector;
- Increased global trade; and

Preserved large land areas from deforestation by intensifying production on more productive lands.

Criticisms of the green revolution

- *Loss of biodiversity:* The spread of Green Revolution hybrids and the associated techniques have resulted in the cultivation of many fewer varieties of crops. Some crops have seen upwards of a 90% reduction in crop varieties. Dependence on one or a few cultivars of a crop means a greater exposure to famine due to a new crop pest, external dependence of the population for other foodstuffs, and an impaired ability to improve crops in the future through breeding.
- *Health value and food quality:* The replacement of multiple staple crops by a single HYV staple crop can mean a less varied diet. In addition, critics argue, many Green Revolution crops are bred for high caloric efficiency, storage longevity, and appearance; but not for health value. As such, many hybrid crops are claimed to be inferior in nutritional value than their ancestors, potentially leading to malnutrition.
- A side-effect of the pesticides used is that the chemicals have killed not only the pests, but also fish in the paddy ecosystem.
- *Health effects:* The chemicals such as insecticides, pesticides and herbicides needed to protect the HYV crops are not only toxic to insects or pests or weeds, but also to humans. People in First World countries may use protection when spraying these chemicals on the plants, but protection is generally not used in Third World countries. Firstly, many farmers are too

poor to buy protective suits. Secondly, many do not trouble to put on protection; wearing such protection while working outdoors in the sun all day can increase the risk of getting heat stroke. As a result, many farmers may be slowly poisoned as their bodies absorb the pesticides and herbicides. 80% of deaths from pesticides occur in the Third World alone.

Globalization and social change

- *Corporate dependence:* Many hybrid strains are sterile, or are sold on the condition that farmers cannot save their seed. F1 hybrids have a much higher yield due to their very high level of heterozygote alleles than their descendants, which makes the propagation of F1-hybrids by farmers less practical. Critics argue that this helps seed companies maximize their profit at the expense of farmers, who are forced to buy new seed each year. Critics have also pointed out that farmers are compelled for competitive reasons to buy hybrid seed, since non-hybrid seeds are not much productive.
- *Social change :* The Green Revolution introduced major changes into a world where the majority of the people still depend on farming for their livelihood. The result of many of these techniques was the encouragement of large-scale industrial agriculture at the expense of small farmers, who were unable to compete with the high-efficiency Green Revolution crops. The result has been massive displacement and increasing urbanization and poverty amongst these farmers, and the loss of their land to large agricultural companies, who are more able to manage the considerable enterprise involved in effectively exploiting Green Revolution techniques.

Sustainability

A final set of criticisms focuses on whether the agricultural practices of the Green Revolution are sustainable.

- *Fossil fuel dependence* : While agricultural output increased as a result of the Green Revolution, the energy input into the processes (that is, the energy that must be expended to produce a crop) has also increased at a greater rate, so that the ratio of crops produced to energy input has decreased over time. Green Revolution techniques also heavily rely on chemical fertilizers, pesticides, and herbicides, some of which must be developed from fossil fuels, making agriculture increasingly reliant on petroleum products. This has raised concerns that a significant decrease in world oil and gas production, and the corresponding price increases.
- *Fertilizer dependence* : Nearly all fertilizers, such as potassium, phosphorus and magnesium, come from limited mineral deposits. (An exception is nitrogen fertilizers, which are produced from inexhaustible atmospheric nitrogen, but which requires methane for production in the Haber process). High-yielding varieties require an increased nutrient input.
- *Pollution* : Fertilizer, pesticide, and herbicide runoff continue to be a significant source of pollution, and a major source of water pollution. Although the dangerous, toxic and sometimes cancer-causing pesticides of the early half of the century (like 2,4,5-T and DDT) have mostly been phased out of agricultural usage (although DDT continues to be used in Third-world nations for control of the *mosquito* which is the transmission vector for malaria), their effects have often not been erased.
- *Land degradation* : Critics charge that the Green Revolution destroys soil quality over the long term. This is a result of a variety of factors, including increased soil salinity that results from heavy irrigation; erosion of the soil, a decreased flux of

organic material to the soil because of lesser allocation of photosynthetical production to stems and roots, and the loss of valuable trace elements. These factors can lead to increased reliance on chemical inputs to compensate for deteriorating soil fertility, a process which may ultimately fail.

Problems of green revolution

Yet, the Green Revolution brought about its own problems. It bypassed million of resource-poor farmers and had unintended social and environmental effects. Some political economists (Cleaver, 1972; Fatam, 1972 and Griffin, 1974) have argued that the Green Revolution technology tended to be monopolized by large commercial farmers who had more access to new information and better economic situation. These farmers gained large profit enabling them to enlarge their landholding by consolidating the farms of small non-adopters through purchase or tenant eviction. This resulted in polarization of rural communities into large commercial farms and the increase in the number of landless farmers.

IntensiveTillage

Conventional agriculture has long been based on the practice of cultivating the soil completely, deeply and regularly. The purpose of this intensive cultivation is to loosen the soil structure to create better drainage, root growth aeration and easier sowing.

Monoculture

Over the last several decades, farmers have turned increasingly to monoculture *i.e.,* growing of only one crop in a field in a year, often on a very extensive scale. Monoculture allows more efficient use of farm machinery for cultivation, sowing, weed control, and harvest and can create economics of scale with regard to purchase of seeds, fertilizer, and pesticides. Monoculture is a natural outgrowth of an industrial approach to agriculture, where labour inputs are minimized and technology-based inputs are maximized in order to increase productive efficiency. In many parts of the world, monocultural production of export crops has largely replaced traditional polycultural subsistence agriculture. Monoculture techniques mesh well with the other practices of modern agriculture; monoculture tends to favour intensive cultivation, application of inorganic fertilizer, irrigation, chemical control of pests, and specialized plant varieties. The link with chemical pesticides is particularly strong, vast fields of the same plant are more susceptible to devastating attack by specific pests and require protection by pesticides.

Application of synthetic fertilizer

The spectacular increase in yields of the past several decades have been due in large part to the widespread and intensive use of synthetic chemical fertilizers. Worldwide, use of fertilizer increased 10 folds between 1950 and 1992.

Produced in large quantities at relatively low cost using fossil fuels and mined mineral deposits, fertilizers can be applied easily and uniformly to crops to supply them with ample amounts of the most essential plant nutrients. Because they meet plants' nutrient needs for the short term, fertilizers have allowed farmers to ignore long-term soil fertility and the processes by which it is maintained.

The mineral components of synthetic fertilizers, however, are easily leached out of the soil. In irrigated systems, the leaching problem may be particularly acute, a large amount of the fertilizer applied to fields actually ends up in streams, lakes, and rivers, where it causes eutrophication. Fertilizer can also be leached into ground water used for drinking, where it poses a significant health hazard. Furthermore, the cost of fertilizer is variable over which farmers have no control since it rises with increases in the cost of petroleum.

Irrigation

An adequate supply of water is the limiting factor for food production in many parts of the world. Thus supplying water to fields from underground aquifers, reservoirs and diverted rivers has been key to increasing overall yield and the amount of land that can be farmed. Although only 16% of the world's agricultural land is irrigated, this land produces 40% of the world's food (Serageldin, 1995). Unfortunately, agriculture is such a prodigious user of water that in many areas where land is irrigated for farming, irrigation has a significant effect on regional hydrology. One problem is that ground water is often pumped faster than it is renewed by rainfall. This overdraft can cause land subsidence, and near the coast it can lead to salt water intrusion. In addition, over drafting of groundwater is essentially borrowing water from the future. Where water for irrigation is drawn from rivers, agriculture is often competing for water with water-dependent wildlife and urban areas. Where dams have been built it hold water supplies, there are usually dramatic effects downstream on the ecology of rivers. Irrigation has another type of impact as well it increase the likelihood that fertilizers will be leached from fields and into local streams and rivers, and it can greatly increase the rate of soil erosion.

Chemical pest and weed control

After World War-II, chemical pesticides were widely treated as the new, scientific weapon in humankind's war against plant pests and pathogens. These chemical agents had the appeal of offering farmers a way to rid their fields once and for all of organisms that continually threatened their crops and literally ate up their profits. But this promise has proved to be false. Pesticides can dramatically lower pest populations in the short term, but because they also kill pests natural predators, pest populations can often quickly rebound and reach even greater numbers than before. The farmer is then forced to use even more of the chemical agents. The dependence on pesticide use that results has been called the "pesticide treadmill". Augmenting the dependence problem is the phenomenon of increased resistance; pest populations continually exposed to pesticides are subjected to intense natural selection for pesticide resistance. When resistance among the pests increases, farmers are forced to apply larger amounts of pesticide or to use different pesticides, further contributing to the conditions that promote even greater resistance.

Although the problem with pesticide dependence may be widely recognized, many farmers especially those in developing nations don't use other options. Global sales of pesticides have continued an upward trend, reaching a record $25 billion in 1994. Ironically, total crop losses to pests have remained fairly constant despite increasing pesticide use (Pimentel *et al.*, 1991).

Besides costing farmers a great deal of money, pesticide including herbicides and fungicides can have a profound effect on the environment and often on human health. Pesticides applied to fields have been easily washed and leached into surface water and groundwater, where they enter the food chain, affecting animal populations at every level and often persisting for decades.

The ways in which conventional agriculture harms future productivity are many. Agricultural resources such as soil, water and genetic diversity are overdrawn and degraded, global ecological processes on which agriculture ultimately depends are altered, and the social conditions conducive to resource conservation are weakened and dismantled (Gilessmann, 1997).

Environmental Issues of Immediate Concern

Soil degradation

Over the past 50 years or so, 1.2 billion ha of land in the world, an area larger than that of China and India taken together has been degraded and its productivity reduced (WRI. 1992). It is also estimated that 500 billion tonnes of top soil has been lost since 1972 (Brown *et. al.,* 1993) and 5 million ha are lost annually due to desertification. If such human-caused losses continue, feeding the world population which is projected to nearly double by the middle of the 21st century will be a very difficult task. Soil degradation is wide spread in India. It is estimated that of the country's total geographical area of nearly 329 million ha. about 188 million ha (57 per cent) is degraded due to various reasons such as wind and water erosion, waterlogging, salinity and desertification whereas in 1947 the area affected by soil degradation was about 110 million ha (TOI, 1997b). With more than half of its land being degraded, India has very bleak prospects of sustaining even the existing (low) crop yield rates in the coming decades if necessary measures to arrest this trend and restore the degraded lands are not taken up on a sufficiently large scale.

Global assessment of soil degradation (GLASOD) defines soil degradation as "a process that describes human induced phenomena which lower the current and/or future capacity of the soil to support human life". The degradation is classified as-light, severe and extreme.

GLASOD was sponsored by UNEP and coordinated by the International Soil Reference and Information Centre (ISRIC) in the Netherlands.

On the basis of studies reported by GLASOD about 750 m.ha of land come under "Light" degradation means loss of some productivity but can be restored through farm conservation practices, the biotic functions are still not very much distorted.

ISRIC (1992) published a world map showing areas of serious concern. Areas of some concern, areas of stable terrain and Non-vegetated land (W.R.I. 1992-93).

On the 1.2 billion hectares with moderate severe and extreme soil degradation adding 750 m ha as "Light degradation" the total degraded level of 1.9 billion hectares constitute 17% of the total vegetated land of the world. The five causes of soil degradation are as follows;

- Vegetation removal entails the removal of vegetation cover
- Over exploitation is the decrease in soil cover through removal of vegetation for fuel wood, fencing and so on.
- Over grazing by livestock leads to a decrease in vegetative cover and trampling of soil.
- Agricultural activities include in sufficient or excessive use of manure and fertilizer; cultivation in steep slopes or in arid areas without proper anti erosion measures, improper irrigation and use of heavy machinery on soils with weak structural stability.
- Industrial and bioindustrial activities result in soils contaminated with pollutants for example through waste discharge, overuse of pesticides and excessive fertilization.

GLASOD in their report informed that vast majority of strongly degraded land is located in Asia (453 m.ha) and Africa (321 m.ha) where poverty and hunger concentrated. If no restoration of 910 m.ha of moderately degraded land is accomplished soon one may fear that at least a part of it may become strongly degraded in the near future. Should the moderately degraded land an area equal to the size of U.S.A. become seriously or extremely degraded, it could have a major impact on the production of food, live stock and forest products. Productivity losses on moderately degraded agricultural land can be mitigated by the increased application of fertilizers. But increasing chemical inputs alone will not reverse the degradation process and may in fact cause serious environmental effects associated with the contamination of surface and ground water.

Modern technologies with fertilizers, pesticides mechanisation are more productive on good soils than on poor soils. Although technology often sustains yields, it only temporarily masks the effects of soil degradation; the yield increases might have been greater if the soil had not been degraded. This difference between the actual and the potential yield represents of income. If soil degradation say from erosion is not corrected the loss of crop yield increase year after year to a point when fertilisation makes no impact but total loss.

According to a 1991 United Nations study, 38% of the land cultivated today has been damaged to some degree by agricultural practices since World War-II (Oldeman *et al.*, 1991). Degradation of soil can involve salting, waterlogging, compaction, contamination by pesticides, decline in the quality of soil structure, loss of fertility, and erosion. Although all these forms of soil degradation are severe problems, erosion is the most widespread. Soil is lost to wind and water erosion at the rate of 5-10 tons per hectare per year in Africa, South America and North America and almost 30 tons per hectare annually in Asia. In comparison, soil is created at the rate of about 1 ton per hectare per year, which means that in just a short period, humans have wasted soil resources that took thousands of years to be built up.

The cause-effect relationship between conventional agriculture and soil erosion is direct and unambiguous. Intensive tillage, combined with monoculture and short rotations, leaves the soil exposed to the erosive effects of wind and rain. The soil lost throughout this process is rich in organic matter, the most valuable soil component. Similarly, irrigation is a direct cause of much water erosion of agricultural soil.

Combined, soil erosion and the other forms of soil degradation render much of the agricultural soil of the world increasingly less fertile. Some land severely eroded or too salty from evaporated irrigation water is lost from production altogether. The land that can still produce is kept productive by the artificial means of adding synthetic fertilizers. Although fertilizers can temporarily replace lost nutrients, they cannot rebuild soil fertility and restore soil health, moreover, their use has a number of negative consequences, as discussed above.

Land degradation

Out of the available 329 million ha of land in the country, nearly 29.22 million ha (10 per cent) is subjected to land degradation through soil fertility loss, water logging, ravines, salinity, alkalinity, shifting cultivation, natural calamities etc. The distribution of the estimated area subject to soil erosion, degradation and natural calamities are presented in Table.

Water logging

Nearly 8.53 million ha of land is subjected to serious water logging problem. Water logging results in restriction of the normal circulation of air inside the soil. When the water table comes to 2 m and above, problem of logging would be felt. Immediately after the monsoon rains vast tract to land is subjected to surface flooding and the problem recedes by June of next year. In major command areas covering 9.63 million ha in 37 major irrigation projects situated in 15 states, water logging is felt in 0.74 million ha.

Salt affected soils: Salt affected soils are of two types: saline soils and alkaline soils: and groups of salt affected soils.

Electrolyte	Types of Salt Affected Soil	Environment	Main Adverse Effect
NaCl & SO_4	Saline soil	Arid & Semi-arid	High osmotic pressure/ toxic effect
Na ions capable of alkaline hydrolysis	Alkaline soil	Semi-arid, semi-humid and humid.	Alkali pH effect on physical properties.

Saline ground water, high water table, ingress of sea and irrigation without the provision of drainage results in salinization and these mainly occurs in arid, semi-arid and coastal areas. As per 1984-85 statistic, 5.50 million ha of land is subjected to soil salinity. The alkali soils of 3.58 million ha occur in Indo-Gangetic plains and parts of Madhya Pradesh.

Ravine lands : Four million ha of land is affected by gully erosion in Gujarat, Madhya Pradesh, Rajasthan and Uttar Pradesh.

Desertification is a major environmental problem that has a direct impact on the lives of people in terms of decreased food out put, depletion of natural resources and deterioration of the environment. Desertification is a process whereby the productivity of land decreases because of a variety of factors, including deforestation, over cultivation, over grazing, poor irrigation management, soil erosion and other practices that affect soil health indirectly under tremendous pressure with highly competing demands of rising population and livestock.

Causes of desertification

- Unsuitable agricultural practices such as extensive cropping over grazing and shifting cultivation without adequate period of recovery.

- Conversion of prime forestland to non-forestland and agricultural land for other uses.
- Salinization and water logging due to poor management of irrigation water.
- Industrial and mining activities without satisfactory measures for land reclamation.

Irrigated land occupies about three per cent of the world's drylands, rainfed cropland about 9 per cent and range land about 88 per cent. Approximately one quarter of the irrigated land, half of the rainfed cropland and three quarters of range lands are desertified. Asia has the highest percentage of its irrigated land desertified. Africa carries that distinction for rainfed cropland.

Shifting cultivation practiced by about 250 million people world over in about 350 m ha is a major cause of declining forest cover. Deforestation accentuates the carbon losses to atmosphere and results in huge losses of soil and soil fertility particularly in short cycle. Overgrazing by herds is far larger than what the land can sustain, year after year has progressively rendered them marginal or wastelands.

Desertification of Drylands

The united Nations Conference on Desertification held in Nairobi, Kenya, in 1977 defined desertification as the diminution or destruction of the biological potential of the land, and can lead ultimately to desert like conditions. Drought was considered as just one cause of desertification. More importantly, human misuse of the land or mismanagement practices like overgrazing, tree cutting, improper tillage methods, poor water distribution systems and overexploitation, are now regarded as the direct cause of desertification.

The principal desertification processes are (i) degradation of vegetative cover, (ii) water erosion, (iii) wind erosion,

(iv) salinization, (v) soil crusting and compaction, (vi) reduction in soil organic matter and (vii) excess of toxic substances in the soil (Dregne, 1990). The list suggests that the world land degradation in an arid land context, is nothing but desertification.

Degradation on drylands (arid, semiarid, and sub-humid) is called desertification because desert like conditions appear where none existed before. Desertified drylands are characterized by low primary productivity (less than 400 kilograms of dry matter per hectare per year) and poor rain use efficiency (usually less than 1 kilogram of dry matter produced per hectare per year for each millimeter of rainfall (World Resources, 1988-89).

Estimates of soil erosion world wide are rough. Crop lands lose far more top soil each year than is replaced naturally. About six to sixteen million hectares of crop land are rendered unproductive because of soil erosion each year. Sodification (water logging, salinization and alkalinization) is responsible for reducing the productivity of upto 1.5 million hectare of crop land each year.

Drylands cover 18 per cent of the land area in developing countries, highest proportion of drylands are in Central America (28 per cent) and Africa (25 per cent). More than 300 million people in developing countries live in drylands.

Table: Degraded Land Distribution

Region	Cause of Degradation	Area degraded m ha
Europe	Industrial	19.7
N. America	Agricultural activity	62.0
Africa	Overgrazing	242.0
Central America	Agricultural Activity	28.0
Asia	Deforestation	298.0
Oceania	Overgrazing	82.0
S. America	Deforestation	99.0

(As on 1990) WR. 92-93

Table: Causes of Soil Degradation (Per cent)

Regional	D.f	O.E.	O.G.	A.A.	Ind/Bio
Europe (219)	38	—	23	29	9
N. America (95)	-	4	30	66	—
Africa (494)	14	13	49	24	—
C. America (63)	22	18	15	45	—
Oceania (103)	12	—	80	8	—
Asia (747)	40	6	26	27	—
S. America (243)	41	5	28	26	—
World (1964)	30	7	35	20	1

W.R.I. 1992-93 m.ha of the area.

D.f. = Deforestation. O.E. Over exploitation. O.G. = Overgrazing.

A.A. = Agricultural activities

Ind = Industrial/Bio-industrial

Table: Types of Soil Degradation (as 5 of vegetated land)

Region country	Water Erosion	Wind Erosion	Chemical degradation	Physical degradation
Europe	52	19	12	17
N. America	63	36	-	1
Africa	46	38	12	4
C. America	74	7	11	8
Oceania	81	16	1	2
Asia	58	30	10	2
S. America	51	17	29	3
World	56	28	12	4
	(1093)	(548)	(239)	(83) = (1963)

Table: Human Induced Soil Degradation

Total Vegetated land (m.ha)	11,553
Total Degraded area (m.ha)	1964
Classification of 1964 m.ha degraded area	
Wind erosion	548.3 (28)
Water erosion	1093.7 (56)
Chemical degradation	239.0 (12)
Physical degradation	83.3 (4)
Total	**1964 (100)**
Wind Erosion	548
Top soil	454 (83)
Terrain deformation	82 (15)
Over blowing	12 (2)
Total	**548 (100)**
Water Erosion	1094
Top soil	920 (54)
Terrain deformation	174 (15)
Total	**1094 (100)**
Chemical Degradation	239
Salinisation	76 (32)
Pollution	22 (9)
Acidification	6(3)
Total	**239 (100)**
Physical Degradation	83
Compaction	68 (82)
Waterlogging	10 (12)
Subsidence of Organic soils	5(6)
Total	**83 (100)**

W.R.I. 92-93 Washington U.S.A.

On the basis of data presented by International Soil Reference and Information Centre (ISRIC) Wageningine, the Netherlands, 1990, the following main conclusions may be drawn. For the world an area of 1.2 billion hectares (approximately the size of China and India combined) suffered moderate to extreme soil degradation caused mainly by agricultural activities, deforestation, overgrazing in the past 45 years. This area represents 11% of the Earth's vegetated surface.

The four major types of degradation are water erosion, wind erosion, chemical degradation and physical and for the world these are set at 56, 28 and 4% of the total vegetated land respectively.

Relating to causes of soil degradation the five recognised are Deforestation, over exploitation, overgrazing, agricultural activities and industrialization. The degree of severity from each for the world is 30, 7, 35, 28 and 1, respectively. But the region that suffered most from each area Asia and Central America, Oceania, North America and Europe respectively.

The impact of human activities over the past half a century when the world population doubled in two directions (a) land degradation to a point that their biotic functions are damaged and (b) deforestation increased by 50% during the 80's to an average of nearly 17 m ha per year, leading to soil erosion to an extent of loss of 6 million hectares of arable land. For India, water erosion affects 90 million hectares and deforestation floods annually about 15 million hectares and to the washing of 2.4 billion tonnes of silt into the rivers. FAO reports that environmental degradation is increasing at a pace that is impairing the productivity of land and undermining the welfare of hundreds of millions of rural population.

Recent studies on soil degradation in various parts of the world under four categories Light, Moderate, Severe and Extreme would be of interest.

Water and overuse of water

Depletion of groundwater resources

Water is essential for survival of all forms of life on this planet earth. Adequate and timely availability of water for irrigation is an important factor affecting agricultural production and thereby food security. The global renewable water resources are estimated at 41.022 cubic kilometers (cubic km) and in 1998 the per capita availability of water was 6.918 cubic meters with wide variations from nation to nation ranging from merely 11 cubic meters in Kuwait to 606.498 cubic meters in Iceland (WRI, 1998). Due to increasing area being brought under irrigation, growing industrialisation and urbanisation and increasing human and animal populations, pressure on a water resources and competition for capturing them have increased tremendously and consequently both surface and groundwater resources are being depleted and degraded at a fast rate in most of the countries of the world. Some people believe that in the 21st century, there will be more water wars both internationally and intra-nationally than any other kinds of war. This trend poses a real threat to not only sustainable development but also human survival.

India as a whole is reasonably well endowed with water resources with the average per capita availability of 1.896 cubic meters of renewable water annually which is higher than the average quantity available in many other countries of the world (WRI, 1998). But there are wide variations in the availability of water across space and over time due to highly uneven distribution of rainfall. Further more, mismanagement of both surface and groundwater resources in conjunction with the growing demand of water for agricultural, industrial and domestic uses has engineered many problems such as depletion and degradation of groundwater aquifers, waterlogging, salinity, pollution of surface water bodies and acute shortage of freshwater in arid, semi-arid hard rock areas in the country. Groundwater table has gone down drastically in many areas of the country such as Mehsana district in north Gujarat and Coimbatore district in west Tamil Nadu. It is estimated that in Mehsana district, water table has been falling at the rate of 5-8 meters annually and that some 2,000 wells dry up every year (Moench and Kumar, 1997). In the coastal areas of Gujarat, excessive extraction has depleted the groundwater aquifers and the vacuum so created has been filled in by intrusion of sea water - a phenomenon called salinity ingress. It is estimated that salinity ingress is increasing at an alarming rate of one-half to one km a year along 60 per cent of the 1000 km long Saurashtra coast (TOI, 1998). The salinity ingress has rendered groundwater in those areas unfit for both domestic and agricultural uses and has adversely affected crop yields.

Most of the problems in the use and management of water resources could be traced to lack of well defined property rights and absence of appropriate institutions for regulating the use of water. Restoring the sustainability of renewable water resources requires governmental intervention in such forms as creation and enforcement of appropriate property rights, regulation of water use, credit, subsidies, taxes and cooperative management by users groups or association (Singh, 1993a).

Fresh water is becoming increasingly scarce in many parts of the world as industry, expanding cities and agriculture compete for limited supplies. Some countries have too little water for any additional agricultural or industrial development to occur. To meet demands for water in many other places, water is being drawn from underground aquifers much faster than it can be replenished by rainfall, and rivers are being drained of their water to the detriment of aquatic and riparian ecosystems and their dependent wildlife.

Agriculture accounts for approximately two third of global water use and is a leading cause of regional water shortages. Agriculture uses so much water in part because it uses water waste-fully. More than half of the water applied to crops is never taken up by the plants it is intended for (Van Tuijl, 1993). Instead, this water either evaporates or drains out of fields. Some wastage of water is inevitable, but a great deal of waste could be eliminated if agricultural practices were oriented toward conservation of water rather than maximization of production. For example, crop plants could be watered with drip irrigation systems, and production of water intensive crops such as rice could be shifted away from regions with limited water supplies.

In addition to using too large a share of the world's fresh water, conventional agriculture has an impact on regional and global hydrological patterns. By drawing such large quantities of water from natural reservoirs on land, agriculture has caused a massive transfer of water from the continents to the oceans. A 1994 study concluded that this transfer of water involves about 190 billion cubic meters of water annually and has raised sea level by an estimated. 1.1 centimeters (Sahagian *et al.,* 1994). Regionally, where irrigation is practiced on a large scale, agriculture brings about changes in hydrology and microclimate. Water is transferred from natural water courses to fields and the soil below them, and increased evaporation changes humidity levels and may affect rainfall patterns. These changes in turn significantly impact natural ecosystems and wildlife.

If conventional agriculture continues to use water in the same ways, regional water crises will become increasingly common, either short-changing the environment, marginalized people and future generations, or limiting irrigation-dependent food production.

Loss of genetic diversity

Throughout most of the history of agriculture, humans have increased the genetic diversity of crop plants worldwide. We have been able to do this both by selecting for a variety of specific and often locally adapted traits through plant breeding and by continually recruiting wild species and their genes into the pool of domesticated plants. In the last several decades, however, the overall genetic diversity of domesticated plants has declined. Many varieties have become extinct, and a great many others are heading in that direction. In the meantime, the genetic base of most major crops has become increasingly uniform. Only six varieties of corn, for example, account for more than 70% of the world's corn crop.

The loss of genetic diversity has occurred mainly because of conventional agriculture's emphasis on short-term productivity gains. When highly productive varieties are developed.

Reduced biodiversity

There is a strong interdependent link between degraded habitats and loss of flora and fauna. Degradation of habitats as a result of agricultural practices leads to a decrease in biological diversity. Failure to maintain biodiversity upsets the ecological balance and may lead to other problems such as weed and pest infestation. Farming practices may have an impact on biodiversity in a variety of ways, including:

- clearing land of native flora and fauna, which may also lead to weed invasion
- removal of natural competitors/ predators from an ecosystem by spraying crops with pesticides, which may lead to plagues of pest and weed species
- addition of plant fertilisers leading, through subsequent run-off, to eutrophication.

To conserve biodiversity, both species and habitat diversity need to be maintained. Farm planning and good management practices are important for maintaining species diversity.

Extinction represents an irrevocable and highly regrettable loss of a portion of the biodiversity of earth. Extinction can be a natural process, caused by random catastrophic events, biological interactions such as competition, disease and predation, chronic stresses or frequent disturbance. In modern times, however, humans are the dominant force causing extinction, mostly because of over harvesting and habitat destruction. During the last 200 years, a global total of perhaps 100 species of mammals, 160 birds, and many other taxa are known to have become extinct through some human influence, in addition to untold numbers of undescribed, tropical species. The greatest value of biodiversity is yet unknown. Much of the earth's great biodiversity is rapidly disappearing, even before we know what is missing. Most biologists agree that life on earth is now faced with the most severe extinction episode, since the event that drove the dinosaurs to extinction 65 million years ago. Species of plants, animals, fungi and microscopic organisms such as bacteria are being lost at alarming rates – so many, fact, that biologists estimate that three species go extinct every hour.

The biological heritage of the planet earth is increasingly at risk. It is estimated that one-quarter of species are in danger of extinction; and 5,000 to 150,000 species are lost annually due to destruction of biomass and habitat by destructive agriculture, deforestation, population and other destructive flashing and grazing practices (Bartelmus, 1997). Much of the world's biodiversity is found in developing countries and it is estimated to be disappearing at 50 to 100 times natural rates (World Bank, 1998). India is endowed with fairly good natural resources and an immensely rich biodiversity. It is rated as one of the 12 mega-diversity countries in the world accounting for 60-70 per cent of the world's biodiversity. As one of the world's oldest and largest agricultural countries. India has an impressive diversity of crop species and varieties. It has 6 per cent of the world's flowering plant species, 14 per cent of the world's birds, one-third of the world's identified plant species numbering over 45,000 and about 81,000 identified species of animals (World Bank, 1996). At least 16 species of crop plants and 320 species of wild relatives of cultivated crops originate in the subcontinent. About 90 per cent of all medicines in India come from plant species, many of which are harvested in the wild. Medicinal plants and other non-timber forest products are particularly important as a source of income and sustenance for tribal population. Natural ecosystems strongly influence natural resources development and management which is important not only for agriculture but also for industrial and municipal development.

Natural resources including environment and biodiversity in India have been under great biotic pressure for decades now. India's high levels of human and domestic animal populations, their high density and rapid growth, high incidence of poverty, high level of literacy and lack of appropriate institutional frame work have all contributed to degradation of natural resources and environment. There has been a tremendous loss of biodiversity due to deforestation also. Many plant and animal species are on the brink of extinction.

All these indicators of loss of sustainability necessitate radical changes in conventional economic planting and policy making. Generally speaking in response to such threats to sustainability, two new paradigms are emerging "economics" and 'sustainable development'. The former focuses on internationalisation of environmental costs into conventional micro and macro-economics and the latter advocates compliance with social and environmental norms in the processes and activities necessary for economic growth. Economics can be seen as an attempt to accommodate extrernalities in the conventional economic analysis while at the same time incorporating in it the criteria of intergenerational equity declined as long term maintenance of per capita consumption. This implies a shift from GDP maximisation towards more sustainable growth which can be declined as environmentally adjusted Net Domestic Product (Bartelmus, 1997).

Towards sustainable use of energy in agroecosytems

Examining agriculture through the lens of energy reveals a critical source of unsustainability. Conventional agriculture is today using more energy to produce food than the food itself contains, and most of this invested energy comes from sources with a finite supply. We have come to depend on fossil fuels to produce our food, yet fossil fuels will not always be available in abundant supply. Moreover, dependence on fossil fuel use in agriculture is linked with virtually every other source of unsustainability in our food production systems.

Problems with intensive fossil fuel use

Growing levels of energy inputs to agriculture have played an important role in increasing yield levels in many of the world's agricultural ecosystems over the past several decades. Most of these energy inputs come from industrial sources, and most are based on the use of fossil fuels. If the strategy for meeting the food demands of the growing population of the world continues to depend on these sources, several critical problems will begin to emerge. Some of these problems are ecological, but others are economic and social.

Deforestation

Forests are a very valuable renewable natural resource providing the vital life support system on this planet earth. They perform multiple functions including that carbon uptake and sequestration and provide various other economic benefits and environmental amenities of public goods type. Being the repository of biodiversity, they constitute an essential element of sustainable development (Nadkarni, 1996). With the fast growing human and animal populations, industrialisation and urbanisation, the demand on forest resources has been progressively increasing. Consequently, this valuable resource is being degraded and depleted all over the world especially in developing countries. The world's average annual rate of deforestation in tropical countries is estimated at 16.9 million ha. This is 50 per cent higher than the rate estimated in the previous 1980. Tropical Forest Resources Assessment (WRI, 1992). Although, there is no consensus about the extent of deforestation of tropical forests in the world, environmentalists point with alarm to eroding hill sides, barren drylands studded with trunks of once- thriving trees, and burned out tropical forests. Of the three tropical regions, namely, Asia, Africa & Latin America. Asia's deforestation rate is the highest at 1.2 per cent per annum for the period 1981-90.

In India, based on the interpretation of Landsat imageries, it is estimated that out of 75 million ha of area recorded as forest, only 64 million ha sustains the actual forest cover and out of this only 35 million ha has adequate cover, which accounts for only about 11 per cent of the total geographical area of the country at present. The National Forest Policy (1988) stipulates that the country as a whole should aim at keeping about one-third of the geographical area under forest cover. It is estimated that about 30 million ha of the total forest land is deforested and degraded and is in need of restoration and rejuvenation (SWD, 1984). India has taken several measures including social forestry projects and Joint Forest Management projects to afforest/ reforest degraded revenue and forest lands and to regenerate denuded natural forests and project them involving local people. As a result, the rate of deforestation has significantly declined and is currently lower than the reforestation rate; in fact, over the period 1990-96, on an average, the rate of reforestation exceeded the rate of deforestation by 7,200 ha per annum (World Bank, 1999).

Pollution of the environment

More water pollution comes from agriculture than from any other single source. Agricultural pollutants include pesticides, herbicides, other agrochemicals, fertilizers and salts.

Pesticides and herbicides applied in large quantities on a regular basis are easily spread beyond their targets, killing beneficial insects and wildlife directly and poisoning farm workers. The pesticides that make their way into streams, rivers and lakes and eventually the ocean can have serious deleterious effects on aquatic ecosystems. They can also affect other ecosystems indirectly. Fish eating raptors, for example, may eat pesticide-laden fish, reducing the reproductive capacity of the raptors and thereby impacting terrestrial ecosystems. Although persistent organochloride pesticides such as DDT known for their ability to remain in ecosystems for many decades are being used less in many parts of the world, their less persistent replacements are often much more acutely toxic. Pesticides and other agro-chemicals also enter groundwater, where they contaminate drinking water supplies. Pesticide contamination of ground water has occurred in at least 26 states. An EPA study in 1995 found that of 29 cities tested in the Midwest, 28 had herbicides present in their tap water.

Fertilizer leached from fields is less directly toxic than pesticides, but its effects can be equally damaging ecologically. In aquatic and marine ecosystems it promotes the overgrowth of algae, causing eutrophication and the death of many types of organisms. Nitrates from fertilizer are also a major contaminant of drinking water in many areas. Rounding out the list of pollutants from agriculture are salts and sediments which in many locales have degraded streams, helped destroy fisheries and rendered wetlands unfit for bird life.

It is clear that conventional agricultural practices are degrading the environment globally, leading to declines in biodiversity, upsetting the balance of natural ecosystems, and ultimately compromising the natural resource base on which humans and agriculture depend.

Pollution by fertilisers, pesticides and herbicides

The entire part of this topic with slight modifications is reproduced from "Organic farming for sustainable Agriculture" Faculty Paper of Manage (*Courtsey;* Manohari, 1999).

A few things need to be considered concerning the issue of fertiliser, pesticide and herbicide use in agriculture:

Many soils are naturally infertile and require additional nutrients to make them productive. Once crops are harvested the nutrients they have removed from the soil are removed from the farming system, and these nutrients need to be replaced, even in fertile soils.

- Chemicals used wisely can contribute to the health of the soil. For example, a reduction in the use of tillage (as a means of controlling weeds) by the strategic use of suitable herbicides greatly minimises soil erosion and associated pollution of downstream areas and minimises the use of tractors, which have an adverse effect on soil structure.
- It is important that fertiliser use is managed to minimise off-site effects. Examples of management include: matching fertiliser inputs to crop demand; controlling erosion to ensure fertilisers are not removed with soil; and careful use of irrigation, especially where permeable soils may allow leaching of nutrients into groundwater.

Even with these points in mind, however, use of fertilisers to combat increasingly low soil nutrient levels may result in plant nutrients leaching and/or running off into water catchments. This excess nutrient level may cause varying degrees of eutrophication (other causes of excess nutrient levels in waterways include: sewage; phosphates from detergents; waste from abattoirs or feedlots; manure and dead animals and plants).

Eutrophication occurs when algal blooms (caused by increased nutrient levels) reduce water movement and light penetration, and slow the rate of replenishment of dissolved oxygen in the water. Dead plant material starts to accumulate and bacterial activity increases. The bacteria deplete the dissolved oxygen level through respiration. Some aquatic fauna cannot tolerate even small reductions in dissolved oxygen content, therefore a change in fauna species composition may occur. In severe cases oxygen is totally depleted. Conditions are then termed anaerobic and most living things in the water at that point will die.

Use of some pesticides and herbicides may cause pollution of the environment, causing a decline in the natural flora and fauna. In some cases this may also result in the contamination of agricultural products (e.g. beef contamination by Endosulphan, a pesticide used in the cotton industry to kill Heliothis grubs).

A few things must be considered before spraying crops with pesticides and/or herbicides:

- Integrated pest management strategies are important, and farmers must carefully evaluate their options before they elect to spray. Questions to be asked are:

  a) What is the balance between pests and predators in the management area ?

  b) Is it possible to do nothing? That is, will the problem solve itself or cause an acceptable level of damage?

c) What alternative measures are at the farmer's disposal?

d) How can the chemical be applied so that any adverse impact will be minimised?

- Spraying with a single type of herbicide/pesticide may lead to herbicide/pesticide resistance.
- Manufacturers' instructions should be carefully followed when using pesticides or herbicides.

Chemical fertilizers

Consumption of chemical fertilizers has increased tremendously in recent years. Nitrogen, phosphorous and potassium are the primary fertilizer nutrients which are widely used in our country.

Table: Trends in Consumption of chemical fertilizers in India (in lakh tonnes).

Year	Nitrogenous (N)	Phosphatic (P)	Potassic (K)	Total (N+P+K)	
				In lakh tonnes	Per hectare (kg)
1950-51	0.55	0.08	0.06	0.69	NEG
1960-61	2.10	0.53	0.29	2.92	1.90
1970-71	14.87	4.62	2.28	21.77	13.13
1980-81	36.78	12.14	6.24	55.16	31.83
1988-89	72.51	27.21	10.68	110.40	61.30
1989-90	73.86	30.14	11.68	115.68	63.49
1990-91	79.97	32.21	13.24	125.46	67.49
1991-92	80.46	33.21	13.61	127.28	69.84
1992-93	84.27	28.44	8.84	121.55	65.53(E)
1993-94	87.89	26.69	9.08	123.66	66.69(E)
1994-95	95.07	29.32	11.25	135.64	73.12(E)
1995-96	98.23	28.97	11.56	138.76	74.81(E)
1996-97	103.02	29.77	10.29	143.08	76.75
1997-98	117.38	41.09	14.71	173.18	78.3
2001-02	109.2	42.15	15.68	167.02	81.7
2002-03	114.17	44.16	17.07	175.39	92.8
2004-05	104.74	40.18	16.01	160.94	98.0

The total consumption of chemical fertilizers is in increasing trends from 0.69 lakh tonnes in 1950-51 to 173.18 lakh tonnes in 1997-98 to 160.94 in 2004-05. Decreasing trends in total fertilizers consumption is observed from 127.28 lakh tonnes in 1991-92 to 121.55 lakh tonnes in 1992-93 and, further during 2001-02 to 2004-05. The same decreasing trend is observed in Phosphatic and Potassic fertilizer consumption during the same period. In case of nitrogenous fertilizers, continuous increasing trends is observed from 1950-51 to 1997-98. Fertilizer consumption per hectare was negligible (0.5 kg per hectare) in 1951-52 increased to 76.75 kgs per hectare during 1996-97 and 98.0 during 2004-05.

Table: Consumption of pesticides (technical grade material) (in thousand tonnes)

Year	Pesticides
1950-51	2.35
1960-61	8.62
1970-71	24.32
1980-81	45.00
1988-89	75.89

The trend in consumption of pesticide is increasing from 2.35 thousand tonnes to 75.8 thousand tonnes from 1950-51 to 1988-89.

Pesticide usage

Herbicides

The use of herbicides has gradually improved in the country. During the period 1988 to 1997 herbicides had an annual growth rate of 13.70 percent. The share of herbicides in total pesticides consumption has increased from 4.7 per cent in 1988 to 16.3 percent 1997. The use of herbicides is increasing in agriculturally advanced regions of the country due to rising cost of labour and shift of labour from agriculture to other investors. Nearly, 85 percent of herbicides are used on rice, tea and beans (BAMI–1996).

Insecticides

India is predominantly an insecticides market. The most important crops with regard to insecticide usage are cotton and rice which account for about 70 percent of total pesticide consumption in India. During the last 10 years (1988-97), the consumption of insecticide has declined at an annual compound growth rate of 3.88 percent. Organo phosphates dominate the Indian market with about 50 percent share followed by the synthetic pyrethroids(19 percent), organo chlorines (18 percent), carbamates (4 per cent) and bio-pesticides(1 percent). While organo chlorine group of pesticides has been banned and phased out in advanced countries, India still uses some of this products with adverse impact on environment and human health (BAMI, 1996).

Fungicides

At present, fungicides are most commonly used agro-chemicals for growing food crops and vegetables. In India the use of fungicides is most popular in fruits followed by potato, rice, tea and coffee. The use of fungicides has increased significantly (2.31 percent) during 1988-97 and its share in total pesticide consumption has increased from about 18 percent to about 30 percent.

Fertilizers and sustainability of agriculture

For the last three decades, all over the developed world fingers have been raised on fertilizer, particularly nitrogen, as the number one enemy of sustainable agriculture. One should not be surprised to see this happen if the rates of application are too high such as in Netherlands. Nitrogen is a mobile nutrient both in plants and soils and considering the fact that its efficiency (nitrogen taken up by the above ground portion of crop expressed as percentage of that applied) varies from 30–40% in rice to 60–80% in other cereals, a sizeable amount could be added to the environment as ammonia by volatilization from soil surface, nitrous oxide or elemental nitrogen by denitrification which is not restricted to tropical rice regions, but also applies to temperate regions and finally as nitrates by leaching in underground water. The ammonia going in the atmosphere contributes to acid rains, while N_2O is involved in depletion of ozone layer. What one generally overlooks is that in the case of nitrogen fertilizers we recycle atmospheric nitrogen which is the raw material for ammonia and urea manufacture. At least 30–50% of it is converted into human edible food and about one-third is immobilized in the soil and only the rest goes back to atmosphere either as ammonia or as N_2O or N_2 after denitrification of nitrates. Phosphates, which are not so mobile in soil and get fixed in the soil as insoluble compounds could also leach from very light soils and may also move with eroded surface soil to surface waters such as lakes and ponds. Within the European Union, the average load of nitrogen in 1990 ranged from 35 kg N ha^{-1} in Spain to 195 kg N ha^{-1} in the Netherlands and average loads of phosphate ranged from 17 kg P_2O_5 ha^{-1} in Spain to 57 kg P_2O_5 ha^{-1} in the Belgium. So great has been the concern of European researchers regarding pollution caused by fertilizers (and manures) that it has been referred to as a 'chemical time bomb' and to overcome the explosion of this the need for immediate change in the land management amounting to abandonment of part of the agricultural land has been suggested.

Fertilizers

Continuous use of inorganic fertilizers mainly containing major nutrients NPK in large quantities and neglecting organic and bio-fertilizers paved the way for deterioration of soil health and in turn ill effects on plants, human being and cattle. The adverse effects of using fertilizers are explained below.

- *Nitrate pollution :* Nitrogen is applied to the soil as urea (Which is readily hydrolyzed to ammonium), ammonium nitrate or a combination of ammonium and nitrate. About 40-60 percent of applied nitrogen is lost by voltalization run off, de-nitrification and leaching. The nitrate that is leached causes a lot of visible and invisible hazardous effects.

  Visible effects

  i. Plants become succulent and dark green colour thus becoming more susceptible to pests and diseases. Ex. BPH in paddy in most of the paddy growing regions.

  ii. It increases the growth, weakens the stem and brings lodging in crops like paddy. It reduces the quality of the seed.

  Invisible effects

  i. *Pollution of ground water by nitrates :* Excess nitrate moves below the root zone or into the ground water (once the ground water becoming polluted it remaining for extended periods of time) and draining of such water causes or disease called *"Methemoglobinemia"*, where nitrite (reduced form of nitrate) interferes with oxygen carrying capacity of blood.

  ii. *Japanese encephalitis (JE):* Excess use of urea in rice fields promotes the growth and spread of vectors causing of human disease called JE. Children between the age group between 4-14 years are mainly affected.

  iii. Nitrosomine illness is caused by the presence of secondary amines which causes cancer in human beings.

  iv. Feroxyl nitrates, alkyl nitrates, vapours of HNO_3 and nitrate aerosoles causes respiratory illness

v. HNO_3 in aerosols may lead to acid rains causing lot of damage to ecosystem and buildings

vi. Nitrate oxide produced by de-nitrification damages the stratospheric ozone layer.

- *Eutrophication:* This refers to the process of enrichment of surface water bodies with nutrients, addition of plant nutrients particularly P&N to surface water bodies such as lakes, reservoirs and streams result in intense prolification and accumulation of algae and higher aquatic plants in excessive quantities which can result in detrimental changes in water quality and can significantly interfere with man's use of the water resource.
- *Soil acidification and alkalization:* Development of soil acidification and alkalization due to continuos use of acidic (NH_4 Cl, (NH_2) SO_4 etc.) and basic ($NANO_3$, CAN, basic slag etc.) fertilizers causing imbalance in nutrients availability to crops and effecting activities of beneficial micro organisms.
- Iron, aluminium and manganese toxicities in acidic soil and sodium toxicity in alkali soils effect the availability of other nutrients and deteriorate fertility and productivity of soils.
- The continuous application of 'P' fertilizers can result in the build up of trace metal contaminants such as arsenic and cadmium contained in the fertilizer.
- Excessive application of potassic fertilizers decrease vitamin "C" (ascorbic acid) and carotene content in vegetable and fruits.
- Excessive application of chemical fertilizers lead to malnutrition due to degradation of carbohydrates and proteins both qualitatively and quantitatively.
- Excessive application of chemical fertilizers effects physical properties of soil such as infiltration, soil aeration, soil structure and bulk density etc.

Pesticides

Pesticides enter environment mainly by air, water and soil. Pesticides enter air by spray drift or voltalization from soil or water. The entry of pesticides in water is mainly by surface runoff, sediment transport from treated soil, industrial wastes and direct application of pesticides to control acquatic pests. Soil receives pesticides when the pesticides are directly applied besides runoff from plants, rains and dumping of empty containers of pesticides. The challenges posed by pesticide usage are explained below:

- Indiscriminate and defective handling of the pesticides causes environmental pollution and leads to health hazards.
- Pesticides resistance: Consistent use of pesticide to control pests had led to development of resistance among pests and vectors and adverse effect on non target organisms.
- Destruction of beneficial organisms: Continuous use of pesticides had an adverse effect on beneficial organisms like honeybees, pollinators, parasites and predators. At the height of the American boll worm problem in Guntur and Prakasham districts in Andhra Pradesh in 1986 almost all the predaceous bird fauna were totally exterminated. The crisis in cotton cultivation posed by boll worms, white flies etc. leading to total crop loss and eventual frustration and suicides of many farmers in A.P.

Pesticides poisoning

- *Manufacturing Level*: Persons engaged in manufacturing of insecticides are subjected to insecticider exposure. This results in chronic poisoning. The poisoning symptoms of aldrin, dialdrin and endrine are headache, fatigue, loss of appetite, loss of weight and memory.

- *Operating level:* The majority of cases occur in hot and humid field conditions. The reason is that the operators or farmers do not wear protective clothing.
- *Consumer level:* Chlorinated hydro carbons can accumulate in the adipose tissues of man. It is very difficult to ascertain the extent of safety of residue in human beings. However, there are a number of evidences that some forms of wild life are suffering due to bio-magnification of these residues.

Pesticide residues

The widespread use of pesticides provides many possible sources of pesticides in the environment and living organism. Pesticides after application are known to persist on crop produce, soil, water and air with harmful effects on human health and the environment. In India problem of pesticide residue in food has been studied by Indian Council of Medical Research (ICMR), Indian Council of Agricultural Research (ICAR) and other institutions in an isolated manner.

Pesticides Residues in Water

Sampling area	Pesticide	Residue level
1. Ponds in coffee plantations, Chikmagalore, Karnataka	HCH	0.02 – 0.2 ppm
2. Yamuna, Delhi	DDT	2.9 – 21.8 ppm
3. Srinagar, J & K	HCH	2.5 – 73.5 ppm
4. River Khan, near Indore, M.P.	Total HCH	0.05 – 0.39 ppm
5. River Chambal, near Kota, Rajasthan	Total HCH	0.06 – 1.49 ppm
6. Drinking water source around Bhopal	Total HCH	1.58-15.88 ppm
		3.15-34.77 ppm

Source : Handa and Walia (1996)

DDT Contamination in Milk

State	Total samples	Incidence (%)	Sample above tolerance level (%)	Range (ppm)
Punjab	263	97.7	50.6	ND-1.11
Haryana	120	93.7	05.0	ND-0.33
Himachal Pradesh	120	100.0	55.8	0.006-0.75
Uttar Pradesh	240	57.1	10.8	ND-0.652
Madhya Pradesh	240	95.8	21.7	ND-0.36
Maharashtra	299	100.0	74.2	0.02-0.965
Gujarat	120	100.0	70.0	0.015-0.20
Andhra Pradesh	240	96.7	57.1	ND-2.224
Karnataka	203	22.2	17.7	ND-1.079
Kerala	120	95.8	09.2	ND-0.08
Bihar	120	95.8	19.2	ND-0.08
West Bengal	120	35.8	12.5	ND-2.82
All states of India	2205	81.1	36.0	ND-2.224

Source: ICAR PROJECT 1986 - 1999

Values for DDT and BHC in Human Milk

Country	Year	BHC	DDT
Japan	1977	250	1900
USA	1977-78	NA	NA
Canada	1987	34	840
Great Britain	1979-80	220	1900
Germany	1979-81	450	1900
Italy	1985	7	47
Israel	1985	390	2800
Kenya	1983-85	110	6900
China	1982	6600	6200
India	1988-89	750	3700

Figure by which overall intake exceeds the acceptable daily in take (ADI), based on 2 % fat 0.8 litres of milk per day.

Source : Ho1. E.H.1995 Down to Earth. 4(10). 27-31

Destruction of soil microbes spoiling the soil health.

- Minor pests become major ones
- Increase in investment for crop production
- Severe imbalance in ecology

Herbicides

- *Persistence in soil:* The herbicide applied to one crop may persist in the soil at concentration high enough to damage subsequent sensitive crops.
- *Residues in crops:* At Coimbatore, the sorghum grain & stalk showed detectable amount of residues when atrazine was applied at 0.5 and 1.0 kg/ ha which was well below the MRL.
- *Toxicity:* Herbicides like trifluralin were found associated with nitrosamines which are potent carcinogens. However, at 1.12 kg/ha of trifluralin, the top soil layer (15 cm) would contain only 0.006 ppb of nitrosamine and this is too small amount to cause cancer (Witter, 1980).

Fungicides

- *Emergence of resistant strains:* Improper use of systemic fungicides like (carbendazium) resulted in development of resistant strains of different plant pathogens.
- *Health Hazards:* Maneb and Streptocycline caused dermititis and some people working with captan or in fields treated with it showed symptoms of skin irritation and rashes (Sharma Kaur, 1990).
- *Fungicide residues:* If the fungicides are used judiciously they may pose serious residue problems.
- *Non-target effects:* Copper fungicides used for the control of coffee rust resulted in increased occurrence of coffee leaf miner and of spider mites (Panlam *et al.*, 1976).

Beever *et al.*, (1984) reported that the residues of captan when used as spray against Botrytis storage of kiwi fruit were within the acceptable limits when used as per recommended dose. However, the increased number of sprays resulted in more than acceptable limit of residue levels (Manohari, 1999).

Pollution of waters

More than any other aspect of environmental degradation, pollution of surface and underground waters due to fertilizers has received greater attention. Water can be broadly classified into four categories: (i) non-flowing such as lakes and ponds; (ii) flowing such as rivers and canals; (iii) river estuaries and (iv) underground water.

Both nitrogen and phosphorus eutrophication of lakes and ponds leads to excessive growth of aquatic plants (macrophhytes) and algae (phytoplankton), which deplete the water of oxygen and this may lead to death of fishes and other aquatic animal life. Furthermore, algal and decaying algal and aquatic plant tissue can lead to discolouration and bad odour and this affects recreational and aesthetic water uses.

As regards flowing water, phosphorus is the limiting nutrient. When all phosphorus is used, plant growth ceases irrespective of the amount of nitrogen available[24]. Thus only if adequate phosphorus is available, increasing concentration of nitrates will lead to algal and macrophyte growth.

In estuaries, on the other hand, nitrogen controls the growth of algae and macrophytes. In USA, nitrates and phosphates are suspected of causing *hypoxia* of the Dead Zone of Gulf of Mexico. Nutrient-enriched run-off from agriculture is also incriminated in the *pfiesteria* problem that killed a large number of fish in the Chakaspeake Bay, USA. The recommended level of nitrogen is 0.1 to 1 mg l^{-1}, while the recommended level of phosphorus is nearly one-tenth of that of nitrogen.

Underground water is the major source of drinking water in India and even in developed countries like USA, 90% of the rural population depends on it for drinking water. The safe limit or MCL (maximum contamination level) established by the US Environmental Protection Agency is 45 mg NO_3-N or 10 mg N l^{-1}. The European Union has fixed MCL limit at 50 mg NO_3-N or 11 mg N l^{-1}. Levels above this may lead to methaemoglobinaemia or blue baby syndrome. This is due to conversion of haemoglobin to methaemoglobin due to nitrites formed from nitrates ingested with drinking water. Haemoglobin is involved in transport of oxygen in the body, while methaemoglobin cannot and thus the patient suffers from anexia (lack of oxygen). There are some indications that excess nitrates in the human body may react with amines and form nitrosamines that may lead to gastric cancer, but substantive data are missing.

In India the highest rate of fertilizer N is in Punjab, where an increase in shallow well waters from 0.04–6.15 in 1975 to 0.31–13.3 mg NO_3-N l^{-1} in 1988 has been reported. A recent survey in Delhi reported a range of 26–150 mg NO_3-N l^{-1} in shallow well water, which is very high constituting a warning.

Impact of Global Climate Change on Agriculture

Mankind throughout the world depend on food, fiber and forest products. Continuity and security of agricultural and forest production are therefore of paramount importance. Predicted changes in climate could be expected to alter, perhaps significantly, the levels and relative agricultural and forestry production of different nations over the next few decades.

Agriculture and forestry are also likely to influence the rate and magnitude of such change, as they can be both significant sources and sinks of a number of greenhouse gases. Adaptive management strategies therefore need to be formulated and implemented for these sectors, to enable them to both adapt to future environmental change, and to limit greenhouse gas emissions.

Atmosphere

The atmosphere is the envelope of gas surrounding the earth which is for the most part permanently bound to the earth by the gravitational field. It is composed primarily of nitrogen (78 per cent by volume) and oxygen (21 per cent). There are also small amounts of argon, carbon dioxide and water vapor, as well as trace amounts of other gases and particulate matter.

Trace components of the atmosphere can be very important of atmospheric functions. Ozone amounts on average for 2 parts per million of the atmosphere but is more concentrated in the stratosphere. This stratospheric ozone is critical to the existence of terrestrial life on the planet. Particulate matter is another important trace component. Aerosol loading of the atmosphere, as well as changes in the tiny carbon dioxide component of the atmosphere, can be responsible for significant changes in climate.

Impact of climate on soil resources

Global warming would affect soils directly through increased temperatures and changes in precipitation and potential evapo-transpiration (PET) rates. In boreal areas, higher temperatures and extended growing periods would reduce the extent of permafrost. There and elsewhere, increased soil temperatures would generally increase oxidation rates of organic matter. However, higher primary production because of the rise in atmospheric CO_2

concentration would entail larger soil organic matter inputs and so the net effect on organic matter levels (and hence carbon accumulation) could be positive. Increased rainfall in some regions would tend to increase leaching and erosion rates on upland sites and increase organic matter accumulation, leaching and the prevalence of the anaerobic conditions on lowland sites. Salinization could increase in other areas where increased rates of evapo-transpiration are not compensated by increased rainfall or flooding.

A gradual eustatic rise in sea-level of 1 meter over the next century could erode some coastal land. Impede soil drainage on coastal plants and salinize soils further inland than at present. However, the effects will vary considerably from place to place depending on such local or regional factors as sediment supply to estuaries and coasts, concurrent land subsidence of elevation resulting from natural or human induced causes and human interventions to protect land from the effects of a rising sea level.

It is not yet possible to quantify the resulting soil changes given the present uncertainity, about the amounts and rates of global climate change and particularly concerning regional patterns of temperature, precipitation and coastal geomorphic changes. In most cases, changes in soils by direct human action, whether international or not (either on-side or off-site), are far greater than possible direct climate induced effects. Soil management measures designed to optimize the soils sustained productive capacity would therefore be generally adequate to counteract degradation resulting from climate change. Soils of natural areas, or other and with a low intensity of management such as semi-natural forests used for extraction of wood and other products are less readily protected against the effects of climate change. However, such soils, too are threatened less by climate change than by human actions – off-site, such as those leading to acid deposition, or on-site, such as excessive nutrient extraction under very low input agriculture.

Location-specific monitoring and research are needed to understand environmental change processes, identify appropriate responses and estimate their technical ecological and socio-economic feasibility. They should include studies of recent environments which developed when the relative sea-level was rising gradually, worldwide or locally.

The world's soils can be protected against the effects of climate change if, land users;

- Manage their soils to give them a well aerated, freely draining, stable structure.
- Maintain as far as possible a closed vegetation cover over the soil surface.
- Use integrated plant nutrient management systems to balance the input and off take of nutrients over a cropping cycle or over the years and to maintain or increase soil organic matter.

The conclusions from the various studies and approaches can be summarized as follows:

- Agriculture (crops and livestock) is clearly sensitive to climate change in both positive and negative ways.
- Climate change will most likely bring regional shifts in production and increased demand for irrigation; if extreme weather events were to become more frequent, the outlook for agriculture could become much less favorable.
- Projected impacts depend strongly on the severity of the climate change scenarios, as well as possible changes in climate variability.

Unmitigated climate change due to increasing greenhouse gases would have global consequences such as adverse impacts on crop yields and water resources, international food insecurity triggered by drought, flooding of lands caused by sea-level rise, and migration of people due to environmental changes.

In Climate Change and World Food Supply, Rosenzweig *et al.* (1993) contend the Impact of Climate Change on Crop Production would vary from region to region across the globe. In "The Enhanced Greenhouse Effect and Its Agricultural Impact," Pittock (1990) concludes yields of agricultural crops might drop because of a decrease in stratospheric ozone, which would cause an increase in damaging ultraviolet-B radiation (UV-B).

In *Climate Change and World Agriculture*, Parry (1990) examines the Impact of Climate Change on Sea Level in terms of crop yields, food security, and inundation of lands. The author reports that in addition to direct farmland loss from inundation, agriculture could also experience increased costs due to saltwater intrusion into surface water and groundwater in coastal regions. The United Nations Environment Programme (UNEP) Information Unit for Climate Change (IUCC) Fact Sheet 102 (1990c) presents an overview of climate change and sea-level rise.

Corresponding Social and Economic Effects of Climate Change also come into play. For example, sea-level rise threatens the well-being of those living on small islands, in deltaic regions and low-lying coastal zones, and in other marginal environments. UNEP's IUCC Fact Sheet 111 (1990d) provides information on the socioeconomic impacts of climate change.

Pimentel (1993) offers a general review of the impacts of climate change on global agriculture in "Climate Changes and Food Supply." UNEP's IUCC Fact Sheet 101 (1990b) also provides a brief overview of the agricultural impacts of climate change.

Higher temperatures threaten dangerous consequences: drought, disease, floods, lost ecosystems. And from sweltering heat to rising seas, global warming's effects have already begun. But solutions are in sight. We know where most heat-trapping gases come from: power plants and vehicles. And we know how to curb their emissions: modern technologies and stronger laws. NRDC is working to put these fixes in place. By shifting the perception of global warming from abstract threat to pressing reality, and promoting online activism. By pressing businesses to use less energy and build more efficient products. And by fighting for laws that will speed these advances.

The 2001 IPCC Third Assessment Report concluded that the poorest countries would be hardest hit, with reductions in crop yields in most tropical and sub-tropical regions due to decreased water availability, and new or changed insect pest incidence. In Africa and Latin America many rainfed crops are near their maximum temperature tolerance, so that yields are likely to fall sharply for even small climate changes; falls in agricultural productivity of up to 30% over the 21st century are projected. Marine life and the fishing industry will also be severely affected in some places.

Consequences of potential global climate changes on agricultural production

Many scientists hold the position that agricultural shifts are likely. The possible effects proposed are listed below:

The first direct effect is the composition of the earth atmosphere, such the amount of carbon dioxide and ozone. Gases such as methane, nitrogen dioxide and chloroflourocarbon however, are commonly believed not to have any effect on physiological processes. Some *indirect* effects are climate parameters resulting from climate change, such as temperature, insolation, rainfall, and humidity. Other indirect effects include the side effects due to the climatic changes, such as the increase in the sea level, changes in ocean currents, or tornadoes.

All these influences may combine negatively or positively-the assessment of these effects depend on whether one considers annuals crops (cereals and legumes) or herbaceous perennial cultures (fodder, meadows) or other cultures such as vine or fruit trees. The effects are also different depending on the latitude. In temperate countries, effects are found less negative or even rather beneficial, while in tropical and desertic countries they tend to be adverse. Effects also depend on altitude, for example, places at higher altitudes tend to benefit from a warmer temperature.

Climate change induced by increasing greenhouse gases is likely to affect crops differently from region to region. For example, average crop yield is expected to drop down to 50% in Pakistan according to the UKMO scenario whereas corn production in Europe is expected to grow up to 25% in optimum hydric conditions.

More favourable effects on yield tend to depend to a large extent on realization of the potentially beneficial effects of carbon dioxide on crop growth and increase of efficiency in water use. Decrease in potential yields is likely to be caused by shortening of the growing period, decrease in water availability and poor vernalization.

Temperature potential effect on growing period

Duration of crop growth cycles are above all, related to temperature. An increase in temperature will speed up development. In the case of an annual crop, the duration between sowing and harvesting will shorten (for example, the duration in order to harvest corn could shorten between one and four weeks). The shortening of such a cycle would have an adverse effect on productivity because senescence would occur sooner. Temperature changes could also have serious implications for crops and trees that need vernalisation.

Potential effect of atmospheric carbon dioxide on yield

Carbon dioxide could have both positive and negative consequences

CO_2 is expected to have positive physiological effects by increasing the rate of photosynthesis. Currently, the amount of carbon dioxide in the atmosphere is 380 parts per million. In comparison the amount of oxygen, is very much higher, at 21,000. This means that often plants may be starved of carbon dioxide, being outnumbered by the photosynthetic pollutant oxygen. The effects of an increase in carbon dioxide would be higher on C_3 crops (such as wheat) than on C_4 crops (such as maize), because the former is more susceptible to carbon dioxide shortage. Under optimum conditions of temperature and humidity, the yield increase could reach 36%, if the levels of carbon dioxide are doubled.

A higher level of carbon dioxide would also allow plants to close their stomata, or make the opening smaller, reducing the loss of water through transpiration. This is because higher carbon dioxide levels would allow the stomata to be closed without suffering photorespiration, which due to too much oxygen in ratio to carbon dioxide, in the plant cell's chloroplasts. Due to the carbon dioxide starvation mentioned above, the carbon dioxide molecules are outnumbered by oxygen molecules, oxygen often replaces carbon dioxide in the Calvin Cycle first. This not only halts sugar production but destroys existing sugars, badly stunting growth and crop output. Higher levels of carbon dioxide would reduce this likelihood, allowing sugar production to take place without destructive setbacks due to oxygen. This would mean the plant would be able to be able to allow the waste product of photosynthesis, oxygen to remain longer inside the chloroplasts, which would normally exit through the stomata, which is the normal solution to excess carbon dioxide. Allowing the stomata to be closed, and thus the reduction of loss of water decreases the plants need for water.

However, other studies also show a change in harvest quality. The growth improvement in C_3 plants could favor vegetative biomass to grain biomass; thus leading to a decrease in grain yield.

Carbon dioxide is believed by many scientists to be potentially responsible for increase in agricultural production: a 10-15 % increase for wheat and soybean, 8% for corn and rice for a +2Â°C scenario on average. However, these results mask great differences among countries.

Effect on quality

According to the IPCC's TAR, "The importance of climate change impacts on grain and forage quality emerges from new research. For rice, the amylase content of the grain—a major determinant of cooking quality—is increased under elevated CO_2 (Conroy *et al.*, 1994). Cooked rice grain from plants grown in high-CO_2 environments would be firmer than that from today's plants. However, concentrations of iron and zinc, which are important for human nutrition, would be lower (Seneweera and Conroy, 1997). Moreover, the protein content of the grain decreases under combined increases of temperature and CO_2 (Ziska *et al.*, 1997).

More than 100 studies have shown that higher CO_2 levels lead to reduced plant uptake of nitrogen (and a smaller number showing the same for trace elements such as zinc) resulting in crops with lower nutritional value. This would primarily impact on populations in poorer countries less able to compensate by eating more food, more varied diets, or possibly taking supplements.

Reduced nitrogen content in grazing plants has also been shown to reduce animal productivity in sheep, which depend on microbes in their gut to digest plants, which in turn depend on nitrogen intake.

Global warming and water distribution

Global warming would be able to modify the global distribution of water, possibly leading to several effects, both detrimental and beneficial.

Effects of water availability on productivity

Water is one of the major limiting factors in the growth and production of crops worldwide. In spite of better water efficiency use, higher summer temperature and lower summer rainfall caused by global warming is likely to have adverse effects. The intensification of the extremes of the hydrological global cycle, will have consequences such as more frequent drought in northern sub-tropical areas or desertification extension in arid areas, while causing devastating flood in other areas.

In developed areas of the world agriculture, and competing industry and municipal users mine fossil water supplies. In coastal areas, deep water wells also reverse normal ground water flow towards the ocean, leading to saline water intrusion into aquifers. Further increases in usage would force societies to conform ground water usage to actual recharge rates.

Agricultural surfaces and climate changes

Climate change is likely to increase the amount of arable land near the poles by reduction of the amount of frozen lands. Sea levels are expected to get up to one meter higher by 2100, though this projection is disputed. Rise in sea level would result in agricultural land loss, in particular in areas such as South East Asia. Erosion, submergence of shorelines, salinity of the water table due to the increased sea levels, could mainly affect agriculture through inundation of low-lying lands.

Erosion and fertility

With global warming, soil degradation is more likely to occur, and soil fertility would probably be affected by global warming. However, due to the fact that the ratio of carbon to nitrogen is a constant, a doubling of carbon is likely to imply a higher storage of nitrogen in soils as nitrates, thus providing higher fertilizing elements for plants, providing better yields. The average needs for nitrogen could decrease, and give the opportunity of changing often costly fertilisation strategies.

Due to the extremes of climate that would result, the increase in precipitations would probably result in greater risks of erosion, according to the intensity of the rain. The possible evolution of the organic matter in the soil is a highly contested issue: while the increase in the temperature would induce a greater rate in the production of minerals, lessening the soil organic matter content, the atmospheric CO_2 concentration would tend to increase it.

Potential effects of global climate change on pests, diseases and weeds

A very important point to consider is that weeds would undergo the same acceleration of cycle as cultivated crops, and would also benefit from carbonaceous fertilization. Since most weeds are C_3 plants, they are likely to compete even more than now against C_4 crops such as corn. However, on the other hand, some results make it possible to think that weed killers could gain in effectiveness with the temperature increase.

Global warming would cause an increase in rainfall in some areas, which would lead to an increase of atmospheric humidity and the duration of the wet seasons. Combined with higher temperatures, these could favor the development of fungal diseases. Similarly, because of higher temperatures and humidity, there could be an increased pressure from insects and disease vectors.

Shortage in grain production

Between 1996 and 2003, grain production has stabilized slightly over 1800 millions of tons. In 2000, 2001,

2002 and 2003, grain stocks have been dropping, resulting in a global grain harvest that was short of consumption by 93 millions of tons in 2003.

The earth's average temperature has been rising since the late 1970s, with the three warmest years on record coming in the last five years. In 2002, India and the United States suffered sharp harvest reductions because of record temperatures and drought. In 2003, Europe suffered very low rainfall throughout spring and summer, and a record level of heat damaged most crops from the United Kingdom and France in the Western Europe through Ukraine in the East. Bread prices have been rising in several countries in the region.

Increases in agricultural production

Between 2003 and 2004 worldwide wheat production was increased by 13%, oilseed production by 16%, rice production by 3% and cotton production by 23%.

Conclusions

In the long run, the climatic change could affect agriculture in several ways :

- productivity, in terms of quantity and quality of crops
- agricultural practices, through changes of water use (irrigation) and agricultural inputs such as herbicides, insecticides and fertilizers
- environmental effects, in particular in relation of frequency and intensity of soil drainage (leading to nitrogen leaching), soil erosion, reduction of crop diversity
- rural space, through the loss of previously cultivated lands, land speculation, land renunciation, and hydraulic amenities.

There are large uncertainties to uncover, particularly because there is lack of information on many specific local regions, and include the uncertainties on magnitude of climate change, the effects of technological changes on productivity, global food demands, and the numerous possibilities of adaptation.

Most agronomists believe that agricultural production will be mostly affected by the severity and pace of climate change, not so much by gradual trends in climate. If change is gradual, there will be enough time for biota adjustment. Rapid climate change, however, could harm agriculture in many countries, especially those that are already suffering from rather poor soil and climate conditions, because there is less time for optimum natural selection and adoption. The adoption of efficient new techniques tends to be far from obvious. Some believe developed nations are too well-adapted to the present climate. developing nations also would often have extensive social or technical constraints that prevent them from achieving sustainable production.

While solar radiation and rainfall are major climatic resources, climate is also the single main factor behind the variability of agricultural production in developing and developed countries alike.

Global warming may thus have profound effects on agriculture[1]and food security. Crop agriculture, forestry and livestock are directly involved as sources or sinks of GHG, but they are also among the most vulnerable victims of the foreseen changes.

Although there is no consensus on what will happen to agricultural environments and production, and at what pace, the following consequences are generally accepted by the scientific community:

- climate has considerable inertia, and cannot be reversed over a short period of time;

- future scenarios are uncertain & significant reduction of the uncertainties will require a decade or more;
- water vapour concentrations will increase in the lower atmosphere, and global mean precipitation could increase by 1.5 to 2.5 % per 1 °C of global warming;
- sea level rise may reach about 50 cm by 2100.

The response to the changes by organisms is relatively easy to assess at the eco-physiological level. It includes, for instance, shorter crop cycles, CO_2fertilisation, modifications of coastal/deltaic agriculture, modified crop/animal and pest/disease relations.

Major methodological difficulties are associated with the extrapolation of impacts to the global scale. The literature stresses the possibility of a modification of the current crop geopolitical balance, human population movements and increased global insecurity.

Global Warming and Greenhouse Effect

Anthropogenic activities such as burning of fossil fuel, industrial and agricultural activities have resulted in the accumulation of greenhouse gases like CO_2, N_2O, CH_4, O_3, CFC and others. Their presence in the atmosphere reduces the loss of heat from the earth surface to outer surface resulting in greenhouse effect thereby making the globe warmer.

Greenhouse Effect

The Greenhouse effect is a term that refers to a physical property of the Earth's atmosphere. If the Earth had no atmosphere, its average surface temperature would be very low of about 18° C rather than the comfortable 15° C found today. The difference in temperature is due to a suite of gases called greenhouse gases which affect the overall energy balance of the Earth's system by absorbing infrared radiation. In its existing state, the Earth atmosphere system balances absorption of solar radiation by emission of infrared radiation to space. Due to greenhouse gases, the atmosphere absorbs more infrared energy than it re-radiates to space, resulting in a net warming of the Earth-atmosphere system and of surface temperature. This is the Natural Greenhouse Effect. With more greenhouse gases released to the atmosphere due to human activity, more infrared radiation will be trapped in the earth's surface which contributes to the enhanced greenhouse effect.

The greenhouse effect, first discovered by Joseph Fourier in 1824, and first investigated quantitatively by Svante Arrhenius in 1896, is the process by which an atmosphere warms a planet.

Introduction

The degree of the greenhouse effect is dependent primarily on the concentration of greenhouse gases in the planetary atmosphere. The deep and carbon dioxide-rich atmosphere of Venus causes a *runaway greenhouse effect* with surface temperatures hot enough to melt lead, the atmosphere of Earth creates habitable temperatures, and the thin atmosphere of Mars causes a minimal greenhouse effect. The Goldilocks Principle "Venus is too hot, Mars is too cold, and Earth is just right." Earth has an average surface temperature comfortably between the boiling point and freezing point of water, and thus is suitable for our sort of life, cannot be explained by simply suggesting that our planet orbits at just the right distance from the sun to absorb just the right amount of solar radiation. Our moderate temperatures are also the result of having just the right kind of atmosphere. A Venus-type atmosphere would produce hellish, Venus-like conditions on our planet; a Mars atmosphere would leave us shivering in a Martian-type deep freeze.

Atmospheric scientists first used the term 'greenhouse effect' in the early 1800s. At that time, it was used to describe the naturally occurring functions of trace gases in the atmosphere and did not have any negative connotations. It was not until the mid-1950s that the term greenhouse effect was coupled with concern over climate change. And in recent decades, we often hear about the greenhouse effect in somewhat negative terms. The negative concerns are related to the possible impacts of an enhanced greenhouse effect. It is important to remember that without the greenhouse effect, life on earth as we know it would not be possible.

Instead, parts of our atmosphere act as an insulating blanket of just the right thickness, trapping sufficient solar energy to keep the global average temperature in a pleasant range. The Martian blanket is too thin, and the Venusian blanket is way too thick! The 'blanket' here is a collection of atmospheric gases called 'greenhouse gases' based on the idea that the gases also 'trap' heat like the glass walls of a greenhouse do.

These gases, mainly water vapor (H_2O), carbon dioxide (CO_2), methane (CH_2), and nitrous oxide (N_2O), all act as effective global insulators. To understand why, it's important to understand a few basic facts about solar radiation and the structure of atmospheric gases.

Greenhouse Effect, the capacity of certain gases in the atmosphere to trap heat emitted from the Earth's surface, thereby insulating and warming the Earth. Without the thermal blanketing of the natural greenhouse effect, the Earth's climate would be about 33° Celsius degrees (about 59 Fahrenheit degrees) cooler—too cold for most living organisms to survive.

The greenhouse effect has warmed the Earth for over 4 billion years. Now scientists are growing increasingly concerned that human activities may be modifying this natural process, with potentially dangerous consequences. Since the advent of the Industrial Revolution in the 1700s, humans have devised many inventions that burn fossil fuels such as coal, oil, and natural gas. Burning these fossil fuels, as well as other activities such as clearing land for agriculture or urban settlements, releases some of the same gases that trap heat in the atmosphere, including carbon dioxide, methane, and nitrous oxide. These atmospheric gases have risen to levels higher than at any time in the last 420,000 years. As these gases build up in the atmosphere, they trap more heat near the Earth's surface, causing Earth's climate to become warmer than it would naturally.

Scientists call this unnatural heating effect global warming and blame it for an increase in the Earth's surface temperature of about 0.6 Celsius degrees (about 1 Fahrenheit degree) over the last nearly 100 years. Without remedial measures, many scientists fear that global temperatures will rise 1.4 to 5.8 Celsius degrees (2.5 to 10.4 Fahrenheit degrees) by 2100. These warmer temperatures could melt parts of polar ice caps and most mountain glaciers, causing a rise in sea level of up to 1 m (40 in) within a century or two, which would flood coastal regions. Global warming could also affect weather patterns causing, among other problems, prolonged drought or increased flooding in some of the world's leading agricultural regions.

Understanding the greenhouse effect

Although concern over the effect of increasing greenhouse gases is a relatively recent development, scientists have been investigating the greenhouse effect since the early 1800s. French mathematician and physicist Jean Baptiste Joseph Fourier, while exploring how heat is conducted through different materials, was the first to compare the atmosphere to a glass vessel in 1827. Fourier recognized that the air around the planet lets in sunlight, much like a glass roof.

In the 1850s British physicist John Tyndall investigated the transmission of radiant heat through gases and vapors. Tyndall found that nitrogen and oxygen, the two most common gases in the atmosphere, had no heat-absorbing properties. He then went on to measure the absorption of infrared radiation by carbon dioxide and water vapor, publishing his findings in 1863 in a paper titled "On Radiation Through the Earth's Atmosphere".

Swedish chemist Svante August Arrhenius, best known for his Nobel Prize-winning work in electrochemistry, also advanced understanding of the greenhouse effect. In 1896 he calculated that doubling the natural concentrations of carbon dioxide in the atmosphere would increase global temperatures by 4 to 6 Celsius degrees (7 to 11 Fahrenheit degrees), a calculation that is not too far from today's estimates using more sophisticated methods. Arrhenius correctly predicted that when Earth's temperature warms, water vapor evaporation from the oceans increases. The higher concentration of water vapor in the atmosphere would then contribute to the greenhouse effect and global warming.

The predictions about carbon dioxide and its role in global warming set forth by Arrhenius were virtually ignored for over half a century, until scientists began to detect a disturbing change in atmospheric levels of carbon dioxide. In 1957 researchers at the Scripps Institution of Oceanography, based in San Diego, California, began monitoring carbon dioxide levels in the atmosphere from Hawaii's remote Mauna Loa Observatory located 3,000 m (11,000 ft) above sea level. When the study began, carbon dioxide concentrations in the Earth's atmosphere were 315 molecules of gas per million molecules of air (abbreviated parts per million or ppm). Each year carbon dioxide concentrations increased to 323 ppm by 1970 and 335 ppm by 1980. By 1988 atmospheric carbon dioxide had increased to 350 ppm, an 8 percent increase in only 31 years.

As other researchers confirmed these findings, scientific interest in the accumulation of greenhouse gases and their effect on the environment slowly began to grow. In 1988 the World Meteorological Organization and the United Nations Environment Programme established the Intergovernmental Panel on Climate Change (IPCC). The IPCC was the first international collaboration of scientists to assess the scientific, technical, and socioeconomic information related to the risk of human-induced climate change. The IPCC creates periodic assessment reports on advances in scientific understanding of the causes of climate change, its potential impacts, and strategies to control greenhouse gases. The IPCC played a critical role in establishing the United Nations Framework Convention on Climate Change (UNFCCC). The UNFCCC, which provides an international policy framework for addressing climate change issues, was adopted by the United Nations General Assembly in 1992.

Today scientists around the world monitor atmospheric greenhouse gas concentrations and create forecasts about their effects on global temperatures. Air samples from sites spread across the globe are analyzed in laboratories to determine levels of individual greenhouse gases. Sources of greenhouse gases, such as automobiles, factories, and power plants, are monitored directly to determine their emissions. Scientists gather information about climate systems and use this information to create and test computer models that simulate how climate could change in response to changing conditions on the Earth and in the atmosphere. These models act as high-tech crystal balls to project what may happen in the future as greenhouse gas levels rise. Models can only provide approximations, and some of the predictions based on these models often spark controversy within the science community. Nevertheless, the basic concept of global warming is widely accepted by most climate scientists.

How the greenhouse effect works

The natural greenhouse effect

The greenhouse effect results from the interaction between sunlight and the layer of greenhouse gases in the Earth's atmosphere that extends up to 100 km (60 mi) above Earth's surface. Sunlight is composed of a range of radiant energies known as the solar spectrum, which includes visible light, infrared light, gamma rays, X rays, and ultraviolet light. When the Sun's radiation reaches the Earth's atmosphere, some 25 percent of the energy is reflected back into space by clouds and other atmospheric particles. About 20 percent is absorbed in the atmosphere. For instance, gas molecules in the uppermost layers of the atmosphere absorb the Sun's gamma rays and X rays. The Sun's ultraviolet radiation is absorbed by the ozone layer, located 19 to 48 km (12 to 30 m ha) above the Earth's surface.

About 50 percent of the Sun's energy, largely in the form of visible light, passes through the atmosphere to reach the Earth's surface. Soils, plants, and oceans on the Earth's surface absorb about 85 percent of this heat energy, while the rest is reflected back into the atmosphere—most effectively by reflective surfaces such as snow, ice, and sandy deserts. In addition, some of the Sun's radiation that is absorbed by the Earth's surface becomes heat energy in the form of long-wave infrared radiation, and this energy is released back into the atmosphere.

Certain gases in the atmosphere, including water vapor, carbon dioxide, methane, and nitrous oxide, absorb this infrared radiant heat, temporarily preventing it from dispersing into space. As these atmospheric gases warm, they in turn emit infrared radiation in all directions. Some of this heat returns back to Earth to further warm the surface in what is known as the greenhouse effect, and some of this heat is eventually released to space. This heat transfer creates equilibrium between the total amount of heat that reaches the Earth from the Sun and the amount of heat that the Earth radiates out into space. This equilibrium or energy balance the exchange of energy between the Earth's surface, atmosphere, and space is important to maintain a climate that can support a wide variety of life.

The heat-trapping gases in the atmosphere behave like the glass of a greenhouse. They let much of the Sun's rays in, but keep most of that heat from directly escaping. Because of this, they are called greenhouse gases. Without these gases, heat energy absorbed and reflected from the Earth's surface would easily radiate back out to space, leaving the planet with an inhospitable temperature close to –19°C (2°F), instead of the present average surface temperature of 15°C (59°F).

Absorbed by land, oceans, and vegetation at the surface, the visible light is transformed into heat and re-radiates in the form of invisible infrared radiation. If that was all there was to the story, then during the day earth would heat up, but at night, all the accumulated energy would radiate back into space and the planet's surface temperature would fall far below zero very rapidly. The reason this doesn't happen is that earth's atmosphere contains molecules that absorb the heat and re-radiate the heat in all directions. This reduces the heat radiated out to space. Called 'greenhouse gases' because they serve to hold heat in like the glass walls of a greenhouse, these molecules are responsible for the fact that the earth enjoys temperatures suitable for our active and complex biosphere.

To appreciate the importance of the greenhouse gases in creating a climate that helps sustain most forms of life, compare Earth to Mars and Venus. Mars has a thin atmosphere that contains low concentrations of heat-trapping gases. As a result, Mars has a weak greenhouse effect resulting in a largely frozen surface that shows no evidence of life. In contrast, Venus has an atmosphere containing high concentrations of carbon dioxide. This heat-trapping gas prevents heat radiated from the planet's surface from escaping into space, resulting in surface temperatures that average 462°C (864°F) too hot to support life.

Types of greenhouse gases

The greenhouse gases

Earth's atmosphere is primarily composed of nitrogen (78 per cent) and oxygen (21 per cent). These two most common atmospheric gases have chemical structures that restrict absorption of infrared energy. Only the few greenhouse gases, which make up less than 1 percent of the atmosphere, offer the Earth any insulation. Greenhouse gases occur naturally or are manufactured. The most abundant naturally occurring greenhouse gas is water vapor, followed by carbon dioxide, methane, and nitrous oxide. Human-made chemicals that act as greenhouse gases include chlorofluorocarbons (CFCs), hydrochlorofluorocarbons (HCFCs), and hydrofluorocarbons (HFCs).

Water vapor (H_2O) causes about 60% of Earth's naturally-occurring greenhouse effect. Other gases influencing the effect include carbon dioxide (CO_2) (about 26%), methane (CH_4), nitrous oxide (N_2O) and ozone (O_3) (about 8%). Collectively, these gases are known as greenhouse gases. The greenhouse effect due to carbon dioxide is specifically known as the Callendar effect.

Since the 1700s, human activities have substantially increased the levels of greenhouse gases in the atmosphere. Scientists are concerned that expected increases in the concentrations of greenhouse gases will powerfully enhance the atmosphere's capacity to retain infrared radiation, leading to an artificial warming of the Earth's surface.

Water vapor

Water vapor is the most common greenhouse gas in the atmosphere, accounting for about 60 to 70 percent of the natural greenhouse effect. Humans do not have a significant direct impact on water vapor levels in the atmosphere. However, as human activities increase the concentration of other greenhouse gases in the atmosphere (producing warmer temperatures on Earth), the evaporation of oceans, lakes, and rivers, as well as water evaporation from plants, increase and raise the amount of water vapor in the atmosphere.

Carbon dioxide

Carbon dioxide (CO_2) is one of the greenhouse gases. It consists of one carbon atom with an oxygen atom bonded to each side. When its atoms are bonded tightly together, the carbon dioxide molecule can absorb infrared radiation and the molecule starts to vibrate. Eventually, the vibrating molecule will emit the radiation again, and it will likely be absorbed by yet another greenhouse gas molecule. This absorption-emission-absorption cycle serves to keep the heat near the surface, effectively insulating the surface from the cold of space.

Carbon dioxide constantly circulates in the environment through a variety of natural processes known as the carbon cycle. Volcanic eruptions and the decay of plant and animal matter both release carbon dioxide into the atmosphere. In respiration, animals break down food to release the energy required to build and maintain cellular activity. A byproduct of respiration is the formation of carbon dioxide, which is exhaled from animals into the environment. Oceans, lakes, and rivers absorb carbon dioxide from the atmosphere. Through photosynthesis, plants collect carbon dioxide and use it to make their own food, in the process incorporating carbon into new plant tissue and releasing oxygen to the environment as a byproduct.

In order to provide energy to heat buildings, power automobiles, and fuel electricity-producing power plants, humans burn objects that contain carbon, such as the fossil fuels oil, coal, and natural gas; wood or wood products; and some solid wastes. When these products are burned, they release carbon dioxide into the air. In addition, humans cut down huge tracts of trees for lumber or to clear land for farming or building. This process, known as deforestation, can both release the carbon stored in trees and significantly reduce the number of trees available to absorb carbon dioxide.

As a result of these human activities, carbon dioxide in the atmosphere is accumulating faster than the Earth's natural processes can absorb the gas. By analyzing air bubbles trapped in glacier ice that is many centuries old, scientists have determined that carbon dioxide levels in the atmosphere have risen by 31 percent since 1750. And since carbon dioxide increases can remain in the atmosphere for centuries, scientists expect these concentrations to double or triple in the next century if current trends continue.

Recent measurements of carbon dioxide amounts from Mauna Loa observatory show that CO_2 has increased from about 313 ppm (parts per million) in 1960 to about 375 ppm in 2005. The current observed amount of CO_2 exceeds the geological record of CO_2 maxima (~300 ppm) from ice core data (Hansen, J., Climatic Change, 68, 269, 2005). This suggests that the CO_2 production rate from increased industrial activity (automobile use and fossil fuel generation) and other human activities such as land-use changes has overwhelmed the normal feedback control mechanisms. Global climate model calculations indicate that the elevated CO_2 levels are likely to lead to global warming. There has been an observed global average temperature increase of about 0.5°C since 1960 (Science 308, 1431, 2005). There is still some public controversy about the role of human activities and that of CO_2 and other greenhouse gas increases for global warming.

Methane

Many natural processes produce methane, also known as natural gas. Decomposition of carbon-containing substances found in oxygen-free environments, such as wastes in landfills, release methane. Ruminating animals such as cattle and sheep belch methane into the air as a byproduct of digestion. Microorganisms that live in damp soils, such as rice fields, produce methane when they break down organic matter. Methane is also emitted during coal mining and the production and transport of other fossil fuels.

Methane has more than doubled in the atmosphere since 1750, and could double again in the next century. Atmospheric concentrations of methane are far less than carbon dioxide, and methane only stays in the atmosphere for a decade or so. But scientists consider methane an extremely effective heat-trapping gas one molecule of methane is 20 times more efficient at trapping infrared radiation radiated from the Earth's surface than a molecule of carbon dioxide.

Nitrous oxide

Nitrous oxide is released by the burning of fossil fuels, and automobile exhaust is a large source of this gas. In addition, many farmers use nitrogen-containing fertilizers to provide nutrients to their crops. When these fertilizers break down in the soil, they emit nitrous oxide into the air. Plowing fields also releases nitrous oxide.

Since 1750 nitrous oxide has risen by 17 percent in the atmosphere. Although this increase is smaller than for the other greenhouse gases, nitrous oxide traps heat about 300 times more effectively than carbon dioxide and can stay in the atmosphere for a century.

Fluorinated compounds

Some of the most potent greenhouse gases emitted are produced solely by human activities. Fluorinated compounds, including CFCs, HCFCs, and HFCs, are used in a variety of manufacturing processes. For each of these synthetic compounds, one molecule is several thousand times more effective in trapping heat than a single molecule of carbon dioxide.

CFCs, first synthesized in 1928, were widely used in the manufacture of aerosol sprays, blowing agents for foams and packing materials, as solvents, and as refrigerants. Nontoxic and safe to use in most applications, CFCs are harmless in the lower atmosphere. However, in the upper atmosphere, ultraviolet radiation breaks down CFCs, releasing chlorine into the atmosphere. In the mid-1970s, scientists began observing that higher concentrations of chlorine were destroying the ozone layer in the upper atmosphere. Ozone protects the Earth from harmful ultraviolet radiation, which can cause cancer and other damage to plants and animals. Beginning in 1987 with the Montréal Protocol on Substances that Deplete the Ozone Layer, representatives from 47 countries established control measures that limited the consumption of CFCs. By 1992 the Montréal Protocol was amended to completely ban the manufacture and use of CFCs worldwide, except in certain developing countries and for use in special medical processes such as asthma inhalers.

Scientists devised substitutes for CFCs, developing HCFCs and HFCs. Since HCFCs still release ozone-destroying chlorine in the atmosphere, production of this chemical will be phased out by the year 2030, providing scientists some time to develop a new generation of safer, effective chemicals. HFCs, which do not contain chlorine and only remain in the atmosphere for a short time, are now considered the most effective and safest substitute for CFCs.

Other synthetic chemicals

Experts are concerned about other industrial chemicals that may have heat-trapping abilities. In 2000 scientists observed rising concentrations of a previously unreported compound called trifluoromethyl sulphur pentafluoride.

Although present in extremely low concentrations in the environment, the gas still poses a significant threat because it traps heat more effectively than all other known greenhouse gases. The exact sources of the gas, undisputedly produced from industrial processes, still remain uncertain.

Consequences of greenhouse effect

Global warming

Global warming is a term used to describe the trend of increases in the average temperature of the Earth's atmosphere and oceans that has been observed in recent decades. The scientific opinion on climate change, as expressed by the UN Intergovernmental Panel on Climate Change (IPCC) and explicitly endorsed by the national science academies of the G8 nations, is that the average global temperature has risen 0.6 ± 0.2 °C since the late 19th century, and that it is likely that "most of the warming observed over the last 50 years is attributable to human activities". The increased volumes of carbon dioxide and other greenhouse gases released by the burning of fossil fuels, land clearing and agriculture, and other human activities, are the primary sources of human-induced warming. The natural greenhouse effect keeps the Earth about 33° C warmer than it otherwise would be; adding carbon dioxide to a planet's atmosphere, with no other changes, will make that planet's surface warmer.

Temperature change is just one aspect of the broader subject of climate change. Current research is attempting to further illuminate and quantify the processes and factors that can affect temperature change, especially positive and negative feedback mechanisms.

Based on basic science, observational sensitivity studies and the climate models referenced by the IPCC, temperatures may increase by 1.4 to 5.8 °C between 1990 and 2100. This is expected to result in other climate changes including rises in sea level and changes in the amount and pattern of precipitation. Such changes may increase the frequency and intensity of extreme weather events such as floods, droughts, heat waves, and hurricanes, change agricultural yields, cause glacier retreat, reduced summer streamflows, or contribute to biological extinctions. Although warming is expected to affect the number and magnitude of these events, it is difficult to connect any particular event to global warming. Much of the evidence is statistical; a significant increase in certain events which is correlated with warming.

'Global warming' is a specific case of the more general term 'climate change' (which can also refer to cooling, such as in Ice ages). Furthermore, the term is in principle neutral as to the causes, but in common usage, 'global warming' generally implies a human influence. Note, however, that the UNFCCC uses 'climate change' for human caused change and 'climate variability' for non-human caused change. Some organizations use the term 'anthropogenic climate change' for human induced changes.

Climate change

Temperature change is just one aspect of the broader subject of climate change. The scientific opinion on climate change, as expressed by the UN Intergovernmental Panel on Climate Change (IPCC) and explicitly endorsed by the national science academies of the G8 nations, is that the average global temperature has risen 0.6 ± 0.2 °C since the late 19th century, and that it is likely that "most of the warming observed over the last 50 years is attributable to human activities".

Causes of global warming

Most of the increase in global temperatures is due to human activities causing an amplified greenhouse effect. The primary sources of human-induced global warming are carbon dioxide and other greenhouse gases released by the burning of fossil fuels, land clearing, agriculture, and other human activities. The natural greenhouse effect keeps the Earth about 33 °C warmer than it otherwise would be; adding carbon dioxide to a planet's atmosphere, with no other changes, will make that planet's surface warmer.

The effects of anthropogenic greenhouse gases

The scientific consensus on global warming is that the Earth is warming, and that humanity's greenhouse gas emissions are making a significant contribution. This consensus is summarized by the findings of the Intergovernmental Panel on Climate Change (IPCC). In the Third Assessment Report, the IPCC concluded that "most of the warming observed over the last 50 years is attributable to human activities". This position was recently supported by an international group of science academies from the G8 countries and Brazil, People's Republic of China and India.

The global temperature on both land and sea has increased by 0.6 ± 0.2 °C over the past century [6]. At the same time, the volume of atmospheric carbon dioxide has increased from around 280 parts per million in 1800 to around 315 in 1958, 367 in 2000 (a 31% increase over 200 years), and about 380 in 2006. Other greenhouse gas emissions have also increased. Future carbon dioxide levels are expected to continue rising due to ongoing fossil fuel usage, though the actual trajectory will depend on uncertain economic, sociological, technological, and natural developments. The IPCC Special report on emissions scenarios gives a wide range of future carbon dioxide scenarios, ranging from 541 to 970 parts per million by 2100.

Greenhouse gases in the atmosphere

The atmospheric concentrations of carbon dioxide and CH_4 have increased by 31% and 149% respectively above pre-industrial levels since 1750. This is considerably higher than at any time during the last 650,000 years, the period for which reliable data has been extracted from ice cores. From less direct geological evidence it is believed that carbon dioxide values this high were last attained 40 million years ago. About three-quarters of the anthropogenic emissions of carbon dioxide to the atmosphere during the past 20 years is due to fossil fuel burning. The rest is predominantly due to land-use change, especially deforestation.

The longest continuous instrumental measurement of carbon dioxide mixing ratios began in 1958 at Mauna Loa. Since then, the annually averaged value has increased monotonically from 315 ppmv (see the Keeling Curve). The concentration reached 376 ppmv in 2003. South Pole records show similar growth. The monthly measurements display small seasonal oscillations.

Another important greenhouse gas, methane, is produced biologically. Some biological sources are "natural" such as termites and others are attributable to human activity such as agriculture, e.g., rice paddies. Recent evidence suggests that forests may also be a source (RC) (BBC). Note that this is a contribution to the *natural* greenhouse effect, and not to the *anthropogenic* greenhouse effect (Ealert). Also, at higher latitudes afforestation may increase the albedo (due largely to the effects of winter snow); at these latitudes, this results in a net warming effect (Wired).

Sources of greenhouse gas emissions

Anthropogenic CO_2 emissions from fuel combustion - contributions to total CO_2 emissions, 1990. Source: UNFCCC

Globally, the majority of anthropogenic greenhouse gas emissions arise from fuel combustion. The remainder is accounted for largely by "fugitive fuel" (consumed in the production and transport of fuel), emissions from industrial processes (excluding fuel combustion), and agriculture: these contributed 5.8%, 5.2% and 3.3% respectively in 1990. Current figures are broadly comparable.

Around 17% of emissions are accounted for by the combustion of fuel for the generation of electricity.

A small percentage of emissions come from natural and anthropogenic biological sources, with approximately 6.3% derived from agriculturally produced methane and nitrous oxide.

Positive feedback effects, such as the expected release of possibly as much as 70,000 million tonnes of methane from permafrost peat bogs in Siberia, which have started melting due to the rising temperatures, may lead to significant additional sources of greenhouse gas emissions.

Note that anthropogenic emissions of other pollutants - notably sulphate aerosol - exert a cooling effect; this can account for the plateau/cooling seen in the temperature record in the middle of the 20th century, though this may also be due to intervening natural cycles.

Projected increases in temperature

Based on basic science, observational sensitivity studies and the climate models referenced by the IPCC, temperatures may increase by 1.4 to 5.8 °C between 1990 and 2100. This is expected to result in other climate changes including rises in sea level and changes in the amount and pattern of precipitation. Such changes may increase the frequency and intensity of extreme weather events such as floods, droughts, heat waves, and hurricanes, change agricultural yields, cause glacier retreat, reduced summer streamflows, or contribute to biological extinctions. Although warming is expected to affect the number and magnitude of these events, it is difficult to connect any particular event to global warming. Much of the evidence is statistical; a significant increase in certain events which is correlated with warming.

Historical warming of the earth

Two millennia of temperatures according to different reconstructions, each smoothed on a decadal scale. The unsmoothed, annual value for 2004 is also plotted for reference.

Relative to 1860-1900 the global temperature on both land and sea has increased by 0.75 °C. Temperatures in the lower troposphere have increased between 0.12 and 0.22 °C per decade since 1979. Over the past one or two thousand years before 1850, world temperature is believed to have been relatively stable, with various fluctuations, which are possibly local, such as the Medieval Warm Period or the Little Ice Age.

Based on estimates by NASA's Goddard Institute for Space Studies, 2005 was the warmest year since reliable wide-spread instrumental measurements became available in the late 1800s, beating the previous record set in 1998 by a few hundredths of a degree Celsius. Similar estimates prepared by the World Meteorological Organization and the UK's Climatic Research Unit concluded that 2005 was still only the second warmest year behind 1998.

Depending on the time frame, different temperature records are available. These are based on different data sets, with different degrees of precision and reliability. An approximately global instrumental temperature record begins in about 1860; contamination from the urban heat island effect is believed to be small. A longer-term perspective is available from various proxy records for recent millennia; see temperature record of the past 1000 years for a discussion of these records and their differences. The attribution of recent climate change is clearest for the most recent period of the last 50 years, for which the most detailed data is available. Satellite temperature measurements of the tropospheric temperature date from 1979.

Relationship to ozone depletion

Although they are often interlinked in the mass media, the connection between global warming and ozone depletion is not strong. There are four areas of linkage:

- Global warming from carbon dioxide radiative forcing is expected (perhaps somewhat surprisingly) to *cool* the stratosphere. This, in turn, would lead to a relative *increase* in ozone depletion and the frequency of ozone holes.
- Conversely, ozone depletion represents a radiative forcing of the climate system. There are two opposed effects: reduced ozone allows more solar radiation to penetrate, thus warming the troposphere. But a cooler stratosphere emits less long-wave radiation, tending to cool the troposphere. Overall, the cooling dominates: the IPCC concludes that observed stratospheric O_3 losses over the past two decades have caused a negative forcing of the surface-troposphere system of about -0.15 ± 0.10 W/m^2.
- One of the strongest predictions of the greenhouse effect theory is that the stratosphere will cool. However, although this is observed, it is difficult to use it as an attribution of recent climate change. One of the difficulties of this conclusion includes the fact

that warming induced by increased solar radiation would not have this upper cooling effect. However, similar cooling is caused by ozone depletion.

Ozone depleting chemicals are also greenhouse gases, representing 0.34 ± 0.03 W/m², or about 14% of the total radiative forcing from well-mixed greenhouse gases.

Relationship to global dimming

Some scientists now consider that the effects of the recently recognized phenomenon of global dimming (the reduction in sunlight reaching the surface of the planet, possibly due to aerosols) may have masked some of the effect of global warming. If this is so, the indirect aerosol effect is stronger than previously believed, which would imply that the climate sensitivity to greenhouse gases is also stronger. Concerns about the effect of aerosol on the global climate were first researched as part of concerns over global cooling in the 1970s.

Effects of Global Warming

The predicted effects of global warming are many and various, both for the environment and for human life. The primary cause of global warming is increasing carbon dioxide causing increased temperature. From this flow a variety of secondary effects, including sea level rise, impacts on agriculture, reductions in the ozone layer, increased intensity and frequency of extreme weather events, and the spread of disease. In some cases, the effects may already be being experienced, although it is impossible to attribute specific natural phenomena to long-term global warming.

A new study suggests that global warming could threaten one-fourth of the world's plant and vertebrate animal species with extinction by 2050. They concluded, after estimating potential changes to habitats—and the resulting loss of species in 25 biodiversity "hot spots" around the world (Malcolm, 2005).

The ecologically rich hot spots include South Africa's Cape Floristic Region, the Caribbean Basin, and the tropical regions of the Andes Mountains. These territories compose only a small fraction of the planet's land area but contain large numbers of Earth's flora and fauna.

Weather

Scientists predict that during global warming, the northern regions of the Northern Hemisphere will heat up more than other areas of the planet, northern and mountain glaciers will shrink, and less ice will float on northern oceans. Regions that now experience light winter snows may receive no snow at all. In temperate mountains, snowlines will be higher and snowpacks will melt earlier. Growing seasons will be longer in some areas. Winter and nighttime temperatures will tend to rise more than summer and daytime ones.

The warmed world will be generally more humid as a result of more water evaporating from the oceans. Scientists are not sure whether a more humid atmosphere will encourage or discourage further warming. On the one hand, water vapor is a greenhouse gas, and its increased presence should add to the insulating effect. On the other hand, more vapor in the atmosphere will produce more clouds, which reflect sunlight back into space, which should slow the warming process.

Greater humidity will increase rainfall, on average, about 1 percent for each Fahrenheit degree of warming. (Rainfall over the continents has already increased by about 1 per cent in the last 100 years). Storms are expected to be more frequent and more intense. However, water will also evaporate more rapidly from soil, causing it to dry out faster between rains. Some regions might actually become drier than before. Winds will blow harder and perhaps in different patterns. Hurricanes, which gain their force from the evaporation of water, are likely to be more severe. Against the background of warming, some very cold periods will still occur. Weather patterns are expected to be less predictable and more extreme.

Sea Levels

As the atmosphere warms, the surface layer of the ocean warms as well, expanding in volume and thus raising sea level. Warming will also melt much glacier ice, especially around Greenland, further swelling the sea. Sea levels worldwide rose 10 to 25 cm (4 to 10 in) during the 20th century, and IPCC scientists predict a further rise of 9 to 88 cm (4 to 35 in) in the 21st century.

Sea-level changes will complicate life in many coastal regions. A 100-cm (40-in) rise could submerge 6 percent of The Netherlands, 17.5 per cent of Bangladesh, and most or all of many islands. Erosion of cliffs, beaches, and dunes will increase. Storm surges, in which winds locally pile up water and raise the sea, will become more frequent and damaging. As the sea invades the mouths of rivers, flooding from runoff will also increase upstream. Wealthier countries will spend huge amounts of money to protect their shorelines, while poor countries may simply evacuate low-lying coastal regions.

Even a modest rise in sea level will greatly change coastal ecosystems. A 50-cm (20-in) rise will submerge about half of the present coastal wetlands of the United States. New marshes will form in many places, but not where urban areas and developed landscapes block the way. This sea-level rise will cover much of the Florida Everglades.

Agriculture

A warmed globe will probably produce as much food as before, but not necessarily in the same places. Southern Canada, for example, may benefit from more rainfall and a longer growing season. At the same time, the semiarid tropical farmlands in some parts of Africa may become further impoverished. Desert farm regions that bring in irrigation water from distant mountains may suffer if the winter snowpack, which functions as a natural reservoir, melts before the peak growing months. Crops and woodlands may also be afflicted by more insects and plant diseases.

Animals and plants

Animals and plants will find it difficult to escape from or adjust to the effects of warming because humans occupy so much land. Under global warming, animals will tend to migrate toward the poles and up mountainsides towards higher elevations, and plants will shift their ranges, seeking new areas as old habitats grow too warm. In many places, however, human development will prevent this shift. Species that find cities or farmlands blocking their way north or south may die out. Some types of forests, unable to propagate toward the poles fast enough, may disappear.

Human health

Global warming may extend the range of vectors conveying infectious diseases such as malaria. Bluetongue disease in domesticated ruminants associated with mite bites has recently spread to the north Mediterranean region. Hantavirus infection, Crimean-Congo hemorrhagic fever, tularemia and rabies increased in wide areas of Russia during 2004–2005. This was associated with a population explosion of rodents and their predators but may be partially blamed on breakdowns in governmental vaccination and rodent control programs. Similarly, despite the disappearance of malaria in most temperate regions, the indigenous mosquitoes that transmitted it were never eliminated and remain common in some areas. Thus, although temperature is important in the transmission dynamics of malaria, many other factors are influential.

In a warmer world, scientists predict that more people will get sick or die from heat stress, due to less hotter days than to warmer nights (giving the sufferers less relief). Diseases now found in the tropics, transmitted by mosquitoes and other animal hosts, will widen their range as these animal hosts move into regions formerly too cold for them. Today 45 per cent of the world's people live where they might get bitten by a mosquito carrying the parasite that causes malaria; that percentage may increase to 60 percent if temperatures rise. Other tropical diseases may spread similarly, including dengue fever, yellow fever, and encephalitis. Scientists also predict rising incidence of allergies and respiratory diseases as warmer air grows more charged with pollutants, mold spores, and pollens.

Efforts to control greenhouse gases

Due to overwhelming scientific evidence and growing political interest, global warming is currently recognized as an important national and international issue. Since 1992 representatives from over 160 countries have met regularly to discuss how to reduce worldwide greenhouse gas emissions. In 1997 representatives met in Kyôto, Japan, and produced an agreement, known as the Kyôto Protocol, which requires industrialized countries to reduce their emissions by 2012 to an average of 5 percent below 1990 levels. To help countries meet this agreement cost-effectively, negotiators are trying to develop a system in which nations that have no obligations or that have successfully met their reduced emissions obligations could profit by selling or trading their extra emissions quotas to other countries that are struggling to reduce their emissions. Negotiating such detailed emissions trading rules has been a contentious task for the world community since the signing of the Kyôto Protocol. However, many experts expect that as the scientific evidence about the dangers of global warming continues to mount, nations will be motivated to cooperate more effectively to reduce the risks of climate change.

The total consumption of fossil fuels is increasing by about 1 percent per year. No steps currently being taken or under serious discussion will likely prevent global warming in the near future. The challenge today is managing the probable effects while taking steps to prevent detrimental climate changes in the future.

Damage can be curbed locally in various ways. Coastlines can be armored with dikes and barriers to block encroachments of the sea. Alternatively, governments can assist coastal populations in moving to higher ground.

There are two major approaches to slowing the buildup of greenhouse gases. The first is to keep carbon dioxide out of the atmosphere by storing the gas or its carbon component somewhere else, a strategy called carbon sequestration. The second major approach is to reduce the production of greenhouse gases.

Carbon sequestration

The simplest way to sequester carbon is to preserve trees and to plant more. Trees, especially young and fast-growing ones, soak up a great deal of carbon dioxide, break it down in photosynthesis, and store the carbon in new wood. Worldwide, forests are being cut down at an alarming rate, particularly in the tropics. In many areas, there is little regrowth as land loses fertility or is changed to other uses, such as farming or building housing developments. Reforestation could offset these losses and counter part of the greenhouse buildup.

Many companies and governments in the United States, Norway, Brazil, Malaysia, Russia, and Australia have initiated reforestation projects. In Guatemala, the AES Corporation, a U.S.-based electrical company, has joined forces with the World Resources Institute and the relief agency CARE to create community woodlots and to teach local residents about tree-farming practices. The trees planted are expected to absorb up to 58 million tons of carbon dioxide over 40 years.

Carbon dioxide gas can also be sequestered directly. Carbon dioxide has traditionally been injected into oil wells to force more petroleum out of the ground or seafloor. Now it is being injected simply to isolate it underground in oil fields, coal beds, or aquifers. At one natural gas drilling platform off the coast of Norway, carbon dioxide brought to the surface with the natural gas is captured and reinjected into an aquifer from which it cannot escape. The same process can be used to store carbon dioxide released by a power plant, factory, or any large stationary source. Deep ocean waters could also absorb a great deal of carbon dioxide. The feasibility and environmental effects of both these options are now under study by international teams.

International agreements

International cooperation is required for the successful reduction of greenhouse gases. In 1992, at the Earth Summit in Rio de Janeiro, Brazil, 150 countries pledged to confront the problem of greenhouse gases and agreed to meet again to translate these good intentions into a binding treaty.

In 1997 in Japan, 160 nations drafted a much stronger agreement known as the Kyôto Protocol. This treaty, which has not yet been implemented, calls for the 38 industrialized countries that now release the most greenhouse gases to cut their emissions to levels 5 per cent below those of 1990. This reduction is to be achieved no later than 2012. Initially, the United States voluntarily accepted a more ambitious target, promising to reduce emissions to 7 percent below 1990 levels; the European Union, which had wanted a much tougher treaty, committed to 8 percent; and Japan, to 6 percent. The remaining 122 nations, mostly developing nations, were not asked to commit to a reduction in gas emissions.

Possible beneficial effects

NOAA projects that by the 2050s, there will only be 54% of the volume of sea ice than that there were in the 1950s.

Global warming may also have positive effects. Plants form the basis of the biosphere. By means of photosynthesis, they use solar energy to convert water, nutrients, and carbon dioxide into usable biomass. Plant growth can be limited by a number of factors, including soil fertility, water, temperature, and carbon dioxide concentration. Lack of carbon dioxide can induce photorespiration, which can destroy existing sugars. Thus, an increase in temperature and atmospheric carbon dioxide can stimulate plant growth in places where these are the limiting factors. IPCC models predict that higher carbon dioxide concentrations would only spur growth of flora up to a point however, because in many regions the limiting factors are water or nutrients, not temperature or carbon dioxide. Despite the limiting factor of water, an increase in carbon dioxide concentration has the direct effect of increasing the transpiration efficiency of most plants so that they actually produce more net biomass per unit of water used by the plant. Satellite data shows that the productivity of the northern hemisphere has indeed increased from 1982 to 1991. However, more recent studies found that from 1991 to 2002, wide-spread droughts had actually caused a decrease in summer photosynthesis in the mid and high latitudes of the northern hemisphere. Moreover, an increase in the total amount of biomass produced is not necessarily all good, since biodiversity can still decrease even though a smaller number of species are flourishing.

Melting Arctic ice may open the Northwest Passage in summer, which would cut 5,000 nautical miles from shipping routes between Europe and Asia. This would be of particular relevance for supertankers which are too big to fit through the Panama Canal and currently have to go around the tip of South America. According to the Canadian Ice Service, the amount of ice in Canada's eastern Arctic Archipelago decreased by 15 percent between 1969 and 2004.

The greenhouse effect and global warming

The greenhouse effect is the name applied to the process which causes the surface of the Earth to be warmer than it would have been in the absence of an atmosphere.

Global warming is the name given to an expected increase in the magnitude of the greenhouse effect, whereby the surface of the Earth will amost inevitably become hotter than it is now.

❒❒❒

6 AGRO-ECOLOGY AND SUSTAINABLE AGRICULTURE

Concept of Agro-ecology

Ecology and environment

The word environment is derived from the French verb environner, which means "encircle" or "surround". Thus, our environment can be defined as the physical, chemical and biological world that surrounds us, as well as the complex of social and cultural conditions affecting an individual or community. This broad definition includes the natural world and the technological environment, as well as the cultural and social contexts that shape human lives.

The earth is the only known planet habited by man and other life forms, plants and vegetation. Man and nature have lived together and so long as man's wants were in conformity with nature, there was no problem. But unfortunately, man's ambition for limitless enjoyment and comfort has led him towards the exploitation of nature's wealth so indiscriminately and so shamelessly as to reduce nature's capacity for self stabilization. The indiscriminate exploitation of nature over centuries has created numerous environmental problems. Man's voracious appetite for resources and his desire to conquer nature has put him on collision course with environment. The demands of his explosive technological society impose intense stress on the state of equilibrium with the environment.

Ecology is the branch of biological science concerned with the relationships and interactions between living organisms and their physical surroundings or environment. Living organisms and the environment with which they exchange materials and energy together make up an ecosystem, which is the basic unit of ecology. An ecosystem includes biotic components - the living plants and animals and abiotic components the air, water, minerals and soil that constitute the environment. A third and essential component of most natural ecosystems is energy, usually in the form of sunlight.

Familiar examples of land based or terrestrial ecosystems include forests, deserts, jungles and medows. Water based or aquatic ecosystems include streams, rivers, lakes, marshes and estuaries. There is no specific limitation on the size or boundaries of an ecosystem. A small pond can be studied as a separate ecosystem, as can a desert comprising hundreds of square kilometers. Even the entire surface of earth can be viewed as an ecosystem; the term biosphere is often used in this context.

Ecology, is the study of the relationship of plants and animals to their physical and biological environment. The physical environment includes light and heat or solar radiation, moisture, wind, oxygen, carbon dioxide, nutrients in soil, water and atmosphere. The biological environment includes organisms of the same kind as well as other plants and animals.

The term *ecology* was introduced by the German biologist Ernst Heinrich Haeckel in 1866; it is derived from the Greek *oikos* ("household"), sharing the same root word as *economics*. Thus, the term implies the study of the economy of nature. Modern ecology, in part, began with Charles Darwin. In developing his theory of evolution, Darwin stressed the adaptation of organisms to their environment through natural selection.

The earth's biosphere

The thin mantle of life that covers the earth is called the biosphere. Several approaches are used to classify its regions.

Biomes

The broad units of vegetation are called plant formations by European ecologists and *biomes* by North American ecologists. The major difference between the two terms is that biomes include associated animal life. Major biomes, however, go by the name of the dominant forms of plant life.

Influenced by latitude, elevation and associated moisture and temperature regimes, terrestrial biomes vary geographically from the tropics through the arctic and include various types of forest, grassland, shrub land and desert. These biomes also include their associated freshwater communities: streams, lakes, ponds and wetlands. Marine environments, also considered biomes by some ecologists, comprise the open ocean, littoral (shallow water) regions, benthic (bottom) regions, rocky shores, sandy shores, estuaries and associated tidal marshes.

Ecosystems

A more useful way of looking at the terrestrial and aquatic landscapes is to view them as *ecosystems,* a word coined in 1935 by the British plant ecologist Sir Arthur George Tansley to stress the concept of each locale or habitat as an integrated whole. A *system* is a collection of interdependent parts that function as a unit and involve inputs and outputs. The major parts of an ecosystem are the *producers* (green plants), the *consumers* (herbivores and carnivores), the *decomposers* (fungi and bacteria) and the *nonliving*, or abiotic component, consisting of dead organic matter and nutrients in the soil and water. Inputs into the ecosystem are solar energy, water, oxygen, carbon dioxide, nitrogen and other elements and compounds. Outputs from the ecosystem include water, oxygen, carbon dioxide, nutrient losses and the heat released in cellular respiration, or heat of respiration. The major driving force is solar energy.

Ecosystems function with energy flowing in one direction from the sun and through nutrients, which are continuously recycled. Light energy is used by plants, which, by the process of photosynthesis, convert it to chemical energy in the form of carbohydrates and other carbon compounds. This energy is then transferred through the ecosystem by a series of steps that involve eating and being eaten, or what is called a food web. Each step in the transfer of energy involves several trophic, or feeding, levels: plants, herbivores (plant eaters), two or three levels of carnivores (meat eaters) and decomposers. Only a fraction of the energy fixed by plants follows this pathway, known as the *grazing food web.* Plant and animal matter not used in the grazing food chain, such as fallen leaves, twigs, roots, tree trunks and the dead bodies of animals, support the *decomposer food web.* Bacteria, fungi and animals that feed on dead material become the energy source for higher tropic levels that tie into the grazing food web. In this way nature makes maximum use of energy originally fixed by plants.

The number of tropic levels is limited in both types of food webs, because at each transfer a great deal of energy is lost (such as heat of respiration) and is no longer usable or transferable to the next tropic level. Thus, each tropic level contains less energy than the tropic level supporting it. For this reason, as an example, deer or caribou (herbivores) are more abundant than wolves (carnivores).

Energy flow fuels the *biogeochemical,* or nutrient cycles. The cycling of nutrients begins with their release from organic matter by weathering and decomposition in a form that can be picked up by plants. Plants incorporate nutrients available in soil and water and store them in their tissues. The nutrients are transferred from one tropic level to another through the food web. Because most plants and animals go uneaten, nutrients contained in their tissues, after passing through the decomposer food web, are ultimately released by bacterial and fungal decomposition, a process that reduces complex organic compounds nto simple inorganic compounds available for reuse by plants.

Imbalances

Within an ecosystem nutrients are cycled internally. But there are leakages or outputs and these must be balanced by inputs, or the ecosystem will fail to function. Nutrient inputs to the system come from weathering of rocks, from windblown dust and from precipitation, which can carry material to a great distances. Varying quantities of nutrients are carried from terrestrial ecosystems by the movement of water and deposited in aquatic ecosystems and associated lowlands. Erosion and the harvesting of timber and crops remove considerable quantities of nutrients that must be replaced. The failure to do so results in an impoverishment of the ecosystem. This is why agricultural lands must be fertilized.

If inputs of any nutrient greatly exceed outputs, the nutrient cycle in the ecosystem becomes stressed or overloaded, resulting in pollution. Pollution can be considered an input of nutrients exceeding the capability of the ecosystem to process them. Nutrients eroded and leached from agricultural lands, along with sewage and industrial wastes accumulated from urban areas, all drain into streams, rivers, lakes and estuaries. These pollutants destroy plants and animals that cannot tolerate their presence or the changed environmental conditions caused by them; at the same time they favour a few organisms more tolerant to changed conditions. Thus, precipitation filled with sulfur dioxide and oxides of nitrogen from industrial areas converts to weak sulfuric and nitric acids, known as acid rain and falls on large areas of terrestrial and aquatic ecosystems. This upsets acid-base relations in some ecosystems, killing fish and aquatic invertebrates and increasing soil acidity, which reduces forest growth in northern and other ecosystems that lack limestone to neutralize the acid.

Agro-ecology

Agro ecology is an interdisciplinary field of study that applies ecological principles to the design and management of agricultural systems. Agro ecology concentrates on the relationship of agriculture to the biological, economic, political and social systems of the world. Agro ecological principles shift the focus of agriculture from food production alone to wider concerns, such as environmental quality, food safety, the quality of rural life, humane treatment of livestock and conservation of air, soil and water. Agro ecology also studies how agricultural processes and technologies will be impacted by wider environmental problems such as global warming, desertification, or salinization.

Agroecology is a scientific discipline that defines, classifies and studies agricultural systems from an ecological and socioeconomic perspective. It is also considered the scientific foundation of sustainable agriculture as it provides ecological concepts and principles for the analysis, design and management of productive, resource-conserving agricultural systems. Agroecology integrates indigenous knowledge with modern technical knowledge to arrive at environmentally and socially sensitive approaches to agriculture, encompassing not only production goals, but also social equity and ecological sustainability of the system. In contrast to the conventional agronomic approach that focuses on the spread of packaged uniform technologies, agroecology emphasises vital principles such as biodiversity, recycling of nutrients, synergy and interaction among crops, animals, soil, etc. and regeneration and conservation of resources. The particular methods or technologies promoted by agroecologists build upon local skills and are adapted to local agroecological and socioeconomic conditions. The implementation of such agroecological principles within the context of a pro-poor, farmer-centred rural development strategy is essential for healthy, equitable, sustainable and productive systems.

The science of agroecology, which is defined as the application of ecological concepts and principles to the design and management of sustainable agroecosystems, provides a framework to assess the complexity of agroecosystems (Altieri, 1995). The idea of agroecology is to go beyond the use of alternative practices and to develop agroecosystems with the minimal dependence on high agrochemical and energy inputs, emphasizing complex agricultural systems in which ecological interactions and synergisms between biological components provide the mechanisms for the systems to sponsor their own soil fertility, productivity and crop protection (Altieri and Rosset, 1995).

Agroecology has been variously defined as the ecology of agriculture, the study of ecological functions in farming and the marriage of agriculture and ecology. More specifically, "Agroecology is defined as the application of ecological concepts and principles to the design and management of sustainable agroecosystems" (Gliessman, 1998). This concept has captured the imaginations of farmers and academics who are searching for innovative ways to increase productivity and sustainability of agriculture while maintaining an environment that must endure as well as provide quality of life. Short of dealing with the complexity of improving today's food systems, most research projects and university courses – even in agroecology - focus on the narrow components of agricultural production and their immediate environmental impacts. Such focus does not reflect our expanding vision of how ecology can inform the design and management of the total food system, nor does it build on the ecological foundation that has been used in several educational programs to support the development of sustainable agroecosystems. Study of the ecology of food systems can provide insight on how to deal with questions at the systems level and contribute to development of sustainable societies.

Principles of agro-ecology

Agroecology goes beyond a one-dimensional view of agroecosystems- their genetics, agronomy, edaphology and so on, - to embrace an understanding of ecological and social levels of co-evolution, structure and function. Instead of focusing on one particular component of the agro-ecosystem, agroecology emphasized the interrelatedness of all agro-ecosystem components and the complex dynamics of ecological process (Vandermeer, 1995).

Agro-ecosystem an assembly of mutually interacting organisms and their environment in which materials related to crop production are interchanged in a largely cyclical manner. An ecosystem has physical, chemical and biological components along with energy sources and pathways of energy and materials interchange. The interactions between these three components exert considerable influence on a particular ecosystem. In agro-ecosystem, besides the natural components, some plants are introduced to get benefit for humans and their livestock.

Agro-ecocystems are communities of plants and animals interacting with their physical and chemical environments that have been modified by people to produce food, fibre, fuel and other products for human consumption and processing. Agro-ecology is the holistic study of agro-ecosystems, including all environmental and human elements. It focuses on the form, dynamics and functions of their interrelationships and the processes in which they are involved. An area used for agricultural production, e.g. a field, is seen as a complex system in which ecological processes found under natural conditions also occur, *e.g.* nutrient cycling, predator/pray interactions, competition, symbiosis and successional changes. Implicit in agroecological research is the idea that, by understanding these ecological relationships and processes, agroecosystems can be manipulated to improve production and to produce more sustainably, with fewer negative environmental or social impacts and fewer external inputs (Altieri, 1995).

The design of such systems is based on the application of the following ecological principles (Reinjntjes *et al.,* 1992).

1. Enhance recycling of biomass and optimizing nutrient availability and balancing nutrient flow.
2. Securing favourable soil conditions for plant growth, particularly by managing organic matter and enhancing soil biotic activity.
3. Minimizing losses due to flows of solar radiation, air and water by way of microclimate management, water harvesting and soil management through increased soil cover.
4. Species and genetic diversification of the agroecosystem in time and space.

Enhance beneficial biological interactions and synergism among agrobiodiversity components thus resulting in the promotion of key ecological processes and services.

These principles can be applied by way of various techniques and strategies. Each of these will have different effects on productivity, stability and resiliency within the farm system, depending on the local opportunities, resource constraints and, in most cases, on the market. The ultimate goal of agro ecological design is to integrate components so that overall biological efficiency is improved, biodiversity is preserved and the agro ecosystem productivity and its self-sustaining capacity is maintained. The goal is to design a quilt of agroecosystems within a landscape unit, each mimicking the structure and function of natural ecosystems.

The role of agro-ecology

The agriculture of the future must be both sustainable and highly productive if it is to feed the growing human population. This twin challenge means that we cannot simply abandon conventional practices wholesale and return to traditional or indigenous practices. Although traditional agriculture can provide models and practices valuable in developing sustainable agriculture, it cannot produce the amount of food required to supply distant urban centers and global markets because of its focus on meeting local and small-scale needs.

What is called for, then, is a new approach to agriculture and agricultural development that builds on the resource-conserving aspects of traditional, local and small-scale agriculture while at the same time drawing on modern ecological knowledge and methods. This approach is embodied in the science of agroecology, which is defined as the application of ecological concepts and principles to the design and management of sustainable agroecosystems.

Agroecology provides the knowledge and methodology necessary for developing an agriculture that is on the one hand environmentally sound and on the other hand highly productive and economically viable. It opens the door to the development of new paradigms for agriculture, in part because it undercuts the distinction between the production of knowledge and its application. It values the local, empirical knowledge of farmers, the sharing of this knowledge and its application to the common goal of sustainability.

Ecological methods and principles form the foundation of agroecology. They are essential for determining (i) if a particular agricultural practice, input, or management decision is sustainable and (ii) the ecological basis for the functioning of the chosen management strategy over the long term. Once these are known, practices can be developed that reduce purchased external inputs, lessen the impacts of such inputs when they are used and establish a basis for designing systems that help farmers sustain their farms and their farming communities.

Even though an agroecological approach begins by focusing on particular components of a cropping system and the ecology of alternative management strategies, it establishes in the process the basis for much more. Applied more broadly, it can help us examine the historical development of agricultural activities in a region and determine the ecological basis for selecting more sustainable practices adapted to that region. It can also trace the causes of problems that have arisen as a result of unsustainable practices. Even more broadly, an agroecological approach helps us explore the theoretical basis for developing models that can facilitate the design, testing and evaluation of sustainable agro-ecosystems. Ultimately, ecological knowledge of agro-ecosystem sustainability must reshape humanity's approach to growing and raising food in order for sustainable food production to be achieved worldwide.

Advantages of agro-ecology

In contrast, the agroecological approach favoured by increasing numbers of farmers, NGOs and analysts around the world offers several advantages. First, it is an alternative path to agricultural productivity or intensification that relies on local farming knowledge and techniques adjusted to different local conditions, management of diverse on-farm resources and inputs and incorporation of contemporary scientific understanding of biological principles and resources in farming systems. Second, it offers the only practical way to actually restore agricultural lands that have been degraded by conventional agronomic practices. Third, it offers an environmentally sound and affordable way for smallholders to sustainably intensify production in marginal areas. Finally, it has the potential to reverse the anti-peasant biases inherent in strategies that emphasise purchased inputs and machinery, valuing instead the assets that small farmers already possess, including local knowledge and the low opportunity costs for labour that prevail in the regions where they live. Thus it is an approach that is likely to decrease, rather than exacerbate, inequality and also enhance sustainability.

Evolution of Agro-ecology

Although some scientists played a significant role in the early development of sustainable agriculture, almost all scientific disciplines have ignored it, with the notable exception of ecology and agro-ecology. Ecology as a scientific approach has only existed since the late 19th century and agro-ecological research is less than 50 years old. Ecology is concerned with the relationships between organisms (including humans) within ecosystems and with the associated flows of energy and materials. Agro-ecosystems differ from natural ecosystems in that they are partly powered by auxiliary energy sources (fossil fuels, animal and human power), human management has reduced species diversity, the dominant plant and animal species are artificially selected and they are controlled by humans rather than through natural feedback mechanisms. Within the agro-ecological paradigm, the socio-cultural elements are regarded as important because human relationships with agricultural systems are prime determinants. Concern for the whole and for the study of relationships as they exist within their natural environment are features that distinguish ecology and agro-ecology from most other scientific disciplines. Scientists, given a choice, strive for completeness of understanding and the ecological paradigm is one of the few in common use that provides a reasonable opportunity to achieve this goal. Although agro-ecology has been used since its inception as a means to help explain why sustainable systems are successful, agro-ecologists are now having an influence on our perceptions of sustainability. It is now apparent how agro-ecological principles can be used to design sustainable farming systems.

The history of agro-ecology

The two sciences from which agro-ecology is derived are ecology and agronomy. Ecology has been concerned primarily with the study of natural systems, whereas agronomy has dealt with applying the methods of scientific investigation to the practice of agriculture. The boundary between pure science and nature on the one hand and applied science and human endeavor on the other, has kept the two disciplines relatively separate, with agriculture ceded to the domain of agronomy. With a few important exceptions, only recently much attention has been devoted to the ecological analysis of agriculture.

Evolution of the agro-ecology concept

Gliessman (1998) traced the history of agro-ecology to the early part of the past century when people in ecology and agronomy found common interests. The area of crop ecology included scientists exploring where crops were grown and climatic conditions where each of them were best adapted.

An early instance of cross-fertilization between ecology and agronomy occurred in the late 1920s with the development of the field of crop ecology. These publications date back to the linking of the terms ecology and agronomy in Klages' 1928 article "Crop Ecology and Ecological Crop Geography in the Agronomic Curriculum". K.H.W. Klages is credited as one of the first to discuss ecology and agriculture. Gliessman describes how groups of scientists diverged after World War II, with the ecologists giving more focus to experiments in the natural environment and agronomists dedicating their attention to cultivated systems in agriculture. This separation of interests endured until the 1970s when books and articles began to appear using the term agroecology and the concept of the agroecosystem. Crop ecologists were concerned with where crops were grown and the ecological conditions under which they grew best. In the 1930s, crop ecologists actually proposed the term agroecology as the applied ecology of agriculture. However, since ecology was becoming more of an experimental science of natural systems, ecologists left the "applied ecology" of agriculture to agronomists and the term agroecology seems to have been forgotten.

Following world war II, while ecology moved in the pure science direction, agronomy became increasingly results oriented, in part because of the growing mechanization of agriculture and the greater use of agricultural chemicals. Researchers in each field became less likely to see any commonalties between the disciplines and the gulf between them widened.

In the late 1950s, the maturing of the ecosystem concept prompted some renewed interest in crop ecology and some work in what was termed agricultural ecology. The ecosystem concept provided, for the first time and overall framework for examining agriculture from an ecological perspective, although few researchers actually used it in this way.

Through the 1960s and 1970s, interest in applying ecology to agriculture gradually gained momentum with intensification of community and population ecology research, the growing influence of systems-level approaches and the increase in environmental awareness. An important sign of this interest at the international level occurred in 1974 at the first International Congress of Ecology, when a working group developed a report entitled "Analysis of Agro-ecosystems".

As more ecologists in the 1970s began to see agricultural systems as legitimate areas of study and more agronomists saw the value of the ecological perspective, the foundations of agroecology grew more rapidly. By the beginning of the 1980s, agroecology had emerged as a distinct methodology and conceptual frame work for the study of agroecosystems. An important influence during this period came from traditional farming systems in developing countries, which began to be recognized by many researchers as important examples of ecologically based agroecosystem management (*e.g.*, Gliessman 1978a; Gliessman et al., 1981).

The importance of interdisciplinary approaches and incorporation of social systems methodology into research and education is summarized by Dalgaard *et al.* (2002). In their paper, a compelling case is made for broadening the application of ecological principles using a hierarchy of scale to understand complexity in the production system. This approach is similar to that used by Gliessman (1998). Agroecology is suggested as the logical discipline to integrate across disciplines and different levels of scale. Natural science methods can be used to describe the decision-support tools that will inform design of ecologically sound agriculture, while social science methods can be used to integrate human dimensions and help us to understand the total system in a more precise way.

Today, agroecology continues to straddle established boundaries. On the one hand, agro ecology is the study of ecological process in agroecosystems. On the other, it is a change agent for the complex social and ecological shifts that may need to occur in the future to move agriculture to a truly sustainable basis.

Sustainable Agro-ecosystems

In general, sustainable agroecosystems are those that maintain the resource base upon which they depend, rely on a minimum of artificial inputs from outside the farm system, manage pests and diseases through internal regulating mechanisms and are able to recover from the disturbances caused by cultivation and harvest (Edwards *et al.*, 1990, Altieri 1995, Gliessman, 1998).

In fact, data from documented cases show that the correctly managed, agro-ecological systems:

- exhibit more stable levels of total production per unit area over time,
- produce economically favourable rates of return, in both energetic and monetary terms,
- provide a return to labour and other inputs sufficient to provide an acceptable livelihood to small farmers and their families,
- ensure soil protection and conservation and enhance agro-biodiversity.

Ultimately, sustainability is a test of time: an agro-ecosystem that has continued to be productive for a long period of time without degrading its resource base-either locally or elsewhere can be sustainable.

Learning from existing sustainable systems

The process of identifying the elements of sustainability begins with two kinds of existing systems: natural ecosystems and traditional agroecosystems. Both have stood the test of time in terms of maintaining productivity over long period and each offers a different kind of knowledge foundation. Natural ecosystems provide an important reference point for understanding the ecological basis of sustainability; traditional agroecosystems offer abundant examples of actually sustainable agricultural practices as well as insight into how social systems like cultural, political and economical fit into the sustainability equation. Based on the knowledge gained from these systems, agroecological research can devise principles, practices and designs that can be applied in converting unsustainable conventional agroecosystems into sustainable ones.

Natural ecological systems have evolved over centuries to take efficient advantage of natural resources. Interacting plant and animal species survive well together in each given environment, including its climate and soils. They provide a model of survival and relative stability on which we can model modern agroecosystems. Natural systems are essentially local and they are most often biologically diverse.

Properties of Natural Ecosystems Compared with Sustainable and Conventional Agro-ecosystems

	Natural Ecosystems	Sustainable Agroecosystems*	Conventional Agroecosystems*
Production (yield)	low	low/medium	high
Productivity (process)	medium	medium/high	low/medium
Diversity	high	medium	low
Resilience	high	medium	low
Output stability	medium	low/medium	high
Flexibility	high	medium	low
Human displacement of ecological processes	low	medium	high
Reliance on external human inputs	low	medium	high
Autonomy	high	high	low
Sustainability	high	high	low

* Properties given for these systems are most applicable to the farm scale and for the short- to medium-term time frame. From: Gliessman, S.R., 1997.

Natural ecosystems as reference points

Natural ecosystems and conventional agroecosystems are very different. Conventional agroecosystesm are generally more productive but far less diverse than natural systems. And unlike natural systems, conventional agroecosystems are far from self-sustaining. Their productivity can be maintained only large additional inputs of energy and materials from external, human sources, otherwise very quickly degrade to a much less productive level. In every respect, these two types of systems are at opposite ends of a spectrum.

The key to sustainability is to find a compromise between and function of natural ecosystems yet yields a harvest for human use. Such a system is manipulated to a high degree by humans for human ends and is therefore not self-sustaining, but relies on natural processes for maintenance of its productivity. Its resemblance to natural systems allows the system to sustain over the long term human appropriation of its biomass without large subsidies of industrial cultural energy and without detrimental effects on the surrounding environment.

As the terms in the table indicate, sustainable agroecosystems model has high diversity, resilience and autonomy of natural ecosystems. Compared to conventional systems, they have somewhat lower and more variable yields, a reflection of the variation that occurs from year to year in nature. These lower yields, however, are usually more than offset by the advantage gained in reduced dependence on external inputs and an accompanying reduction in adverse environmental impacts.

From this comparison we can derive a general principle: the greater the structural and functional similarity of an agroecosystem to the natural ecosystems in its biogeographic region, the greater the likelihood that the agroecosystem will be sustainable. If this principles holds true, then observable and measurable values for a range of natural ecosystem processes, structures and rates can provide threshold values, structures and rates can provide threshold values, or benchmarks, that describe or delineate the ecological potential for the design and management of agroecosystems in a particular area. It is the task of research to determine how close an agroecosystem needs to be to these benchmark values to be sustainable (Gliessman, 1990b).

Structure and Functioning of Agro-ecosystem

An agricultural system functions with all its inputs and outputs. Inputs include all materials such as fertilizer, irrigation, pesticides etc for proper functioning of the economic production system and outputs are the gains from the system in any form such as grains, leaves, stems, roots etc. For the sustainability of the production system, agroecosystem research considers all aspects of biology of an agricultural system. This broader approach to agricultural systems becomes more useful as production methods, developed through experimental methods, begin to approach the limits of the biological processes. As these limits are approached, it becomes more important to know how biological processes determine yield and to learn how these processes act in determining the biological limits to yields. Cycling of the nutrients through biogeochemical processes must also be considered in order to determine whether these cycles are able to sustain the high yields of intensive agricultural production that is based on the use of chemicals.

Agroecosystem research uses the methods of ecosystem analysis to measure the material and energy entering plant and animal populations and to explain how these inputs affect the physiological processes determining growth and maintenance. A biologically complete agroecosystem, presented in the diagram shows three basic components and a pattern of energy flow that is nearly the same in all agroecosystems. Green plants harvest the solar energy and store within their bodies in the form of organic compounds. The accumulated organic carbon of the compounds can be divided into three categories: (i) coarse materials, such as fodder eaten by livestock; (ii) grains, fruits, vegetables and other edible portions, eaten by humans and livestock; and (iii) dense stems and leaves which may be used for fuel, shelters or utensils. The paths of energy and material flow between plants and the consumers follow simple consistent patterns that are more or less similar in most agricultural systems. Agroecosystems have relatively simple cycle compared to other ecosystems that form complex food webs.

In what ways can we recapture the knowledge developed over centuries of traditional agricultural production knowledge and link these with the efficiencies of natural systems and with new technologies? Gliessman (1998) suggested that, "The greater the structural and functional similarity of an agroecosystem to the natural ecosystems in its biogeographic region, the greater the likelihood that the agroecosystem will be sustainable." One of our challenges in research is to discover how the principles, the design and the functions of natural systems can be used as benchmarks or guides to development of productive, future systems (Gliessman, 1990, 2001). The Land Institute in Kansas provides one example of a research scenario based on the prairie as a model ecosystem (Jackson, 1980; Soule and Piper, 1992). Their national network of 125 scientists is exploring how an understanding of natural systems can be used to inform our search for productive future perennial-based agroecosystems that operate in harmony with the environment and natural resources.

When we focus only on the production sector in agriculture, the analysis of current systems and design of future alternatives is severely constrained. Such focus ignores the large investment in energy and materials that are integral to the processing, transportation and marketing steps in the food chain. In the industrial food system, it is a practical impossibility to reincorporate many of the waste products in this chain back into the production cycle due to distance, cost and logistical complications. A global system may bring us bananas every day of the year, if we can afford them, but it obscures the principal of seasonality in food production in each place and the cycles that are both inherent and efficient in natural systems. Building on principles of ecology and uniqueness of place, agroecology and analysis of agroecosystems can provide methods for broadening the focus to analyzing all components of the food system and how they interact.

Populations and communities

The functional units of an ecosystem are the populations of organisms through which energy and nutrients move. A population is a group of interbreeding organisms of the same kind living in the same place at the same time. Groups of populations within an ecosystem interact in various ways. These interdependent populations of plants and animals make up the community, which encompasses the biotic portion of the ecosystem.

Diversity

The community has certain attributes, among them dominance and species diversity. Dominance results when one or several species control the environmental conditions that influence associated species. In a forest, for example, the dominant species may be one or more species of trees, such as oak or spruce; in a marine community the dominant organisms frequently are animals such as mussels or oysters. Dominance can influence diversity of species in a community because diversity involves not only the number of species in a community, but also how numbers of individual species are apportioned.

The physical nature of a community is evidenced by layering, or *stratification*. In terrestrial communities, stratification is influenced by the growth form of the plants. Simple communities such as grasslands, with little vertical stratification, usually consist of two layers, the ground layer and the herbaceous layer. A forest has upto six layers: ground, herbaceous, low shrub, low tree and high shrub, lower canopy and upper canopy. These strata influence the physical environment and diversity of habitats for wildlife. Vertical stratification of life in aquatic communities, by contrast, is influenced mostly by physical conditions: depth, light, temperature, pressure, salinity, oxygen and carbon dioxide.

Habitat and niche

The community provides the habitat the place where particular plants or animals live. Within the habitat, organisms occupy different niches. A niche is the functional role of a species in a community that is, its occupation, or how it earns its living. For example, the scarlet tanager lives in a deciduous forest habitat. Its niche, in part, is gleaning insects from the canopy foliage. The more a community is stratified, the more finely the habitat is divided into additional niches.

Community interactions

Major influences on population growth involve various population interactions that tie the community together. These include *competition,* both within a species and among species; *predation,* including parasitism; and *coevolution,* or adaptation.

Competition

When a shared resource is in short supply, organisms compete and those that are more successful survive. Within some plant and animal populations, all individuals may share the resources in such a way that none obtains sufficient quantities to survive as adults or to reproduce. Among other plant and animal populations, dominant individuals claim access to the scarce resources and others are excluded. Individual plants tend to claim and hold onto a site until they lose vigor or die. These prevent other individuals from surviving by controlling light, moisture and nutrients in their immediate areas.

Many animals have a highly developed social organization through which resources such as space, food and mates are apportioned among dominant members of the population. Such competitive interactions may involve *social dominance,* in which the dominant individuals exclude subdominant individuals from the resource; or they may involve *territoriality,* in which the dominant individuals divide space into exclusive areas, which they defend. Subdominant or excluded individuals are forced to live in poorer habitats, do without the resource, or leave the area. Many of these animals succumb to starvation, exposure and predation.

Competition among members of different species results in the division of resources in a community. Certain plants, for example, have roots that grow to different depths in the soil. Some have shallow roots that permit them to use moisture and nutrients near the surface. Others growing in the same place have deep roots that are able to exploit moisture and nutrients not available to surface-rooted plants.

Predation

One of the fundamental interactions is predation, or the consumption of one living organism, plant or animal, by another. While it serves to move energy and nutrients through the ecosystem, predation may also regulate population and promote natural selection by weeding the unfit from a population. Thus, a rabbit is a predator on grass, just as the fox is a predator on the rabbit. Predation on plants involves defoliation by grazers and the consumption of seeds and fruits. The abundance of plant predators, or herbivores, directly influences the growth and survival of the carnivores. Thus, predator-prey interactions at one feeding level influence the predator-prey relations at the next feeding level. In some communities, predators may so reduce populations of prey species that a number of competing species can coexist in the same area because none is abundant enough to control the resource. When predators are reduced or removed, however, the dominant species tend to crowd out other competitors, thereby reducing species diversity.

Parasitism

Closely related to predation is parasitism, wherein two organisms live together, one drawing its nourishment at the expense of the other. Parasites, which are smaller than their hosts, include many viruses and bacteria. Because of this dependency relationship, parasites normally do not kill their hosts the way predators do. As a result, hosts and parasites generally co-evolve a mutual tolerance, although parasites may regulate some host populations, lower their reproductive success and modify behavior.

Co-evolution

Co-evolution is the joint evolution of two unrelated species that have a close ecological relationship—that is, the evolution of one species depends in part on the evolution of the other. Co-evolution is also involved in predator-prey relations. Over time, as predators evolve more efficient ways of capturing or consuming prey, the prey evolves ways to escape predation. Plants have acquired such defensive mechanisms as thorns, spines, hard seed-coats and poisonous or ill-tasting sap that deter would-be consumers. Some herbivores are able to breach these defenses and attack the plant. Certain insects, such as the monarch butterfly, can incorporate poisonous substances found in food plants into their own tissues and use them as a defense against predators. Other animals avoid predators by assuming an appearance that blends them into the background or makes them appear part of the surroundings. The chameleon is a well-known example of this interaction. Some animals possessing obnoxious odors or poisons as a defense also have warning colourations, usually bright colors or patterns, that act as further warning signals to potential predators.

Another co-evolutionary relationship is mutualism, in which two or more species depend on one another and cannot live outside such an association. An example of mutualism is mycorrhizae, an obligatory relationship between fungi and certain plant roots. In one group, called ectomycorrhizae, the fungi form a cap or mantle about the rootlets. The fungal hyphae (threads) invade the rootlet and grow between the cell walls as well as extending outward into the soil from the rootlet. The fungi, which include several common woodland mushrooms, depend on the tree for their energy source. In return the fungi aid the tree in obtaining nutrients from the soil and protect the rootlets of the tree from certain diseases. Without the mycorrhizae some groups of trees, such as conifers and oaks, cannot survive and grow. Conversely, the fungi cannot exist without the trees.

Succession and climax communities

Ecosystems are dynamic, in that the populations constituting them do not remain the same. This is reflected in the gradual changes of the vegetational community over time, known as succession. It begins with the colonization of a disturbed area, such as an abandoned crop field or a newly exposed lava flow, by species able to reach and to tolerate the environmental conditions present. Mostly these are opportunistic species that hold on to the site for a variable length of time. Being short-lived and poor competitors, they are eventually replaced by more competitive, longer-lived species such as shrubs and then trees. In aquatic habitats, successional changes of this kind result largely from changes in the physical environment, such as the buildup of silt at the bottom of a pond. As the pond becomes more shallow, it encourages the invasion of floating plants such as pond lilies and other aquatic ferns. The pace at which succession proceeds depends on the competitive abilities of the species involved; tolerance to the environmental conditions brought about by changes in vegetation; the interaction with animals, particularly the grazing herbivores; and fire. Eventually the ecosystem arrives at a point called the *climax*, where further changes take place very slowly and the site is dominated by long-lived, highly competitive species. As succession proceeds, however, the community becomes more stratified, enabling more species of animals to occupy the area. In time, animals characteristic of later stages of succession replace those found in earlier stages.

Food Web, set of interconnected food chains by which energy and materials circulate within an ecosystem. The food web is divided into two broad categories: the grazing web, which typically begins with green plants, algae, or photosynthesizing plankton and the detrital web, which begins with organic debris. These webs are made up of individual food chains. In a grazing web, materials typically pass from plants to plant eaters (herbivores) to flesh eaters (carnivores). In a detrital web, materials pass from plant and animal matter to bacteria and fungi (decomposers), then to detrital feeders (detritivores) and then to their predators (carnivores).

Generally, many interconnections exist within food webs. For example, the fungi that decompose matter in a detrital web may sprout mushrooms that are consumed by squirrels, mice and deer in a grazing web. Robins are omnivores, that is, consumers of both plants and animals and thus are in both detrital and grazing webs. Robins typically feed on earthworms, which are detritivores that feed upon decaying leaves.

Tropic levels

The food web can be viewed not only as a network of chains but also as a series of tropic (nutritional) levels. Green plants, the primary producers of food in most terrestrial food webs, belong to the first tropic level. Herbivores, consumers of green plants, belong to the second tropic level. Carnivores, predators feeding upon the herbivores, belong to the third. Omnivores, consumers of both plants and animals, belong to the second and third. Secondary carnivores, which are predators that feed on predators, belong to the fourth tropic level. As the tropic levels rise, the predators become fewer, larger, fiercer and more agile. At the second and higher levels, decomposers of the available materials function as herbivores or carnivores depending on whether their food is plant or animal material.

Energy flow

Through these series of steps of eating and being eaten, energy flows from one tropic level to another. Green plants or other photosynthesizing organisms use light energy from the sun to manufacture carbohydrates for their own needs. Most of this chemical energy is processed in metabolism and dissipated as heat in respiration. Plants convert the remaining energy to biomass, both above ground as woody and herbaceous tissue and below ground as roots. Ultimately, this material, which is stored energy, is transferred to the second tropic level, which comprises grazing herbivores, decomposers and detrital feeders. Most of the energy assimilated at the second tropic level is again lost as heat in respiration; a fraction becomes new biomass. Organisms in each tropic level pass on as biomass much less energy than they receive. Thus, the more steps between producer and final consumer, the less energy remains available. Seldom are there more than four links, or five levels, in a food web. Eventually, all energy flowing through the tropic levels is dissipated as heat. The process whereby energy loses its capacity to do work is called entropy.

Factors Affecting Agro-ecological Balance

Is it possible to quantify a healthy agroecosystem? This question is worthy of examination, even though as in many of life's worthier pursuits, quantification is a difficult, if not impossible, activity. Even the subjects of sustainability defy precise definition (Keeney, 1997b). Soil quality, a more objective term, has also been difficult to discuss in quantitative terms (Doran *et al.*, 1997), but descriptive soil health terms are more easily obtained from farmers (Harris and Bezdicek, 1994). Because soil quality is central to healthy agroecosystems, it could well prove difficult to obtain even semi-quantitative evaluations of ecosystem health that could be understandable to scientists, policy makers and the public, though the farmer may know rather precisely, in subjective terms that an agroecosystem is healthy.

Major factors affecting ecological balance are deforestation and overgrazing of rangelands, accelerated soil erosion, irrigation related problems, over exploitation of ground water and indiscriminate use of agrochemicals (*Courtesy;* S.R. Reddy, Principles of Agronomy).

(i) Deforestation and over grazing of rangelands

Perennial vegetation such as trees and grasses successfully prevent soil erosion and runoff from fallows. Deforestation and over grazing leads to soil erosion, silting up of reservoirs and soil degradation. Forests influence climate of a region due to their effect on wind direction and hence the rainfall.

Deforestation and overgrazing modifies the climate and the biodiversity besides loss of valuable genetic resources used in breeding programme for developing high yielding cultivars. While some plant and animal species can adapt to the direct or indirect impact of particular agroecosystem, others have been unable to do so.

(ii) Accelerated soil erosion

Accelerated soil erosion is currently a major environmental problems in tropical and subtropical areas, a consequence of population growth and demand for food. When once the vegetative cover is lost, the bare soil is exposed to the vagaries of wind and intensive rains leading to accelerated soil erosion and making the soil unsuitable for crop production. Runoff from arable land contributes to nutrient enrichment (eutrophication) of the water into which it drains. The result is an increased rate of algae growth, increase in turbidity and depletion of oxygen. Aerobic plants and animals die and the rate of decomposition and recycling of nutrients can decline to a level at which a fresh water take may be regarded as virtually dead.

(iii) Irrigation related problems

Poor quality of water is one for the main factors turning good soils into saline or sodic soils. Provision of irrigation without adequate drainage leads to the same problem as that with poor quality water. Many canal irrigated lands have become unproductive due to salt problems and high ground water table. Total area suffering from water logging ranges between 6 and 8.5 M ha while that affected by salinity is around 9 M ha. It is the misuse and the associated poor water use efficiency which lead to depriving the tail enders of the valuable water resources and diminishing the potential of otherwise prime land which would have otherwise benefited.

(iv) Over exploitation of ground water

Water table depth is one of the main factors determining the economics of irrigation. Rapid increase in number of wells has resulted in a situation in which a large number of borewell owners competing to extract water from the limited aquifers, with a concomitant lowering of ground water table.

Deepening of water tables, in areas where less rain water is captured than necessary to offset the withdrawals, is likely to lead ultimately to permanent water deficit situation. Further, over exploitation of ground water aggravates surfacing of harmful flourides and salts. The areas that face such grave situation abound specially in Peninsular India. Sustainable agriculture aims at maximization of the ground water recharge through appropriate interventions favouring minimization of surface runoff.

(v) Chemical fertilizers

Soluble inorganic fertilizers, particularly nitrogen, which have not been taken up by the plants are leached out of the system. Others such as phosphorus and potassium are not so susceptible to leaching losses as is nitrogen, except under abnormal conditions. Over use and abuse of chemical fertilizer harm biological power of the soil. This must be prevented as all nutrient transformations are negotiated by soil microflora. Organic matter is the source of energy to soil microflora and its content is the index of soil health.

(vi) Pesticides and herbicides

Use of pesticides, fungicides and herbicides to cope up with plant protection opened the doors for several problems. Excessive reliance on synthetic chemicals has resulted in problems such as pesticide resistance, resurgence, residues and environmental pollution. Despite large scale use of such chemicals, more pest outbreaks of bollworms and whitefly in cotton, brown plant hopper and leaf folder in rice, pyrilla and borer in sugarcane and pod borer in pigeonpea are mainly due to destruction of natural enemies. Application efficiency is usually low, with only less than 20 per cent chemical applied reaching the target. Pesticide and herbicide residues are disposable into air, soil and water and other non agricultural ecosystems, where they are subjected to transformation. Pesticides and herbicides enter the agro-ecosystems, via green plants either by direct foliar absorption or by uptake from water and soil. If the pesticide or herbicide taken by humans and animals through food or feed is not voided from the system, these are stored in internal organs such as liver. They are highly toxic at relatively low levels. Residual toxicity contributes to elimination of vulnerable species in animals and plants.

(vii) Air pollution

Rapid increase in industrial activity during last 100 years significantly deteriorated the environmental quality by releasing toxic gases that are harmful to both human and animal population. Major pollution gases are CO_2, SO_2 and oxides of nitrogen (green house gases). These will have indirect effect on plant growth by way of increase in air temperature.

(viii) Increase in air temperature and carbon dioxide

Projections into future (1.4 to 5.8° C) are alarming. Presently over India, the warming is observed in post-monsoon and winter seasons with little variation during monsoon season. The expected change in climate for India indicates that the increase in temperature is likely to be less in *kharif* than in *rabi*. The *rabi* rainfall may exhibit larger uncertainty where as *kharif* rainfall is likely to increase by about 10 per cent. Such global climate change will affect agriculture through its direct and indirect affect on crops, cropping systems, livestock, pests, diseases and soil there by threatening food security.

(ix) Green house gasses

Ability of the atmosphere to admit most of the radiation and retard the re-radiation from earth is known as green house effect. This keeps the planet 33° C warmer than it would otherwise be, allowing the earth to sustain life. Over the last 200 years, the greenhouse gasses (primarily carbon dioxide) altered the composition of atmosphere and caused an enhanced greenhouse raise in temperature would be unprecedented for various ecosystems and the implications are of unpredictable dimensions. Frequency of floods, cyclones, droughts, forest fire etc. is likely to increase.

It is estimated that we can only afford to release a limited amount of carbon dioxide into the atmosphere, otherwise we pass the "safe" limits of climate changes it is at this point that climate change can happen so fast that ecosystems will not be able to adapt. A temperature increase of one degree C is the absolute maximum that could be allowed. The amount of carbon that can be released If we continue to burn fossil fuels at the present levels, the safe limit of one degree Celsius will be reached in just 40 years (Ramakrishna *et al.,* 2000).

(x) Depletion of ozone layer

Depletion of ozone in the upper atmosphere is another threat to mankind. There are alarming reports of increase in UV-B radiation, which reduces the reproduction of phytoplankton, the microorganisms at the bottom of the marine food chain that is vital to the survival of marine animals. The UV-B radiation also induces skin cancer and has accumulating effect on evergreen woody plants. In other words, UV-B damage is cumulative and passes on from one generation to the next. The environmental need has always been and continue to be is to stop using all ozone depleting substances as quickly as possible. The goal, thus, is to minimize further ozone loss but also to take steps for the earliest recovery of the ozone layer.

(xi) Acid rain

Suspended air borne pollutants, sulphur dioxide and other gases in the air react with rain droplets to form acids like H_2SO_4 and these droplets fall on the ground as acid rain. Though the concentration is low, it can cause considerable damage to living and non living things on earth. The threat posed to Taj Mahal by the acid rains due to the oil refinery at Mathura and the tanneries in and around Agra is a classical example. Acid rain also causes considerable damage to agriculture. The acidity of rain is largely due to SO_4 rather than NO_3 which is usually neutralized by NH_4 and Ca. There is evidence of reduction in grain yield of wheat due to acid rain (Singh and Agarwal, 1996).

❑❑❑

7 APPROACHES TO SUSTAINABLE AGRICULTURE

Leisa

Sustainable agriculture is an important element of the overall effort to make human activities compatible with the demands of the earth's eco-system. Thus, an understanding of the different approaches to ecological agriculture is necessary if we want to utilise the planet's resources wisely.

While sustainable agriculture is based on long-term goals but not a specific set of farming practices, it is usually accompanied by a reduction of purchased inputs in favor of managing on-farm resources. A good example is reliance on biologically-fixed nitrogen from legumes versus manufactured nitrogen fertilizers. Low-input agriculture is one of several alternative farming systems whose methods are adaptable to sustainable agriculture.

Low-input farming is based on a reduction—but not necessarily elimination—of chemical fertilizers, insecticides, and herbicides. Farmers are adopting these practices primarily to reduce costs, but also because they want to minimize impact on the environment or because they perceive future pesticide regulations.

In a search for information on how to farm with fewer chemicals, it is helpful to examine alternative farming systems in existence that largely exclude chemicals in favor of biological farming practices. Experiences of producers who have successfully practiced these methods are valuable to farmers considering a transition to low-input sustainable agriculture.

Low-input agriculture

The term low-input agriculture has been defined as a production activity that uses synthetic fertilizers or pesticides below rates commonly recommended by the Extension Service. It does not mean elimination of these materials. Yields are maintained through greater emphasis on cultural practices, IPM, and utilization of on-farm resources and management.

LEISA refers to those forms of agriculture that;

- seek to optimise the use of locally available resources by combining the different components of the farm system, *i.e.,* plants, animals, soil, water, climate and people, so that they complement each other and have the greatest possible synergetic effects;
- seek ways of using external inputs only to the extent that they are needed to provide elements that are deficient in the ecosystem and to enhance available biological, physical and human resources. In using external inputs, attention is given mainly to maximum recycling and minimum detrimental impact on the environment.

LEISA does not aim at maximum production of short duration but rather at a stable and adequate production level over the long term (Reijntjes *et al.,* 1992). Although the term "low-input farming" has often been used to describe any system of alternative agriculture, it can be seen that it is distinctly different from organic farming, etc. Nevertheless, any system that reduces purchased chemical inputs can be called low-input farming.

Characteristics of LEISA and LEIA farms

LEISA	Conventional Agriculture/ LEIA
• Farm management which optimises the use of locally available natural and human resources and indigenous technical knowledge to enhance diversity cyclic flow patterns and to build up living soil.	• Farm practices characterised by the use of inorganic fertilizes, low degree of recycling and low degree of optimising nutrient availability.
• Characterised by a conscious drive towards sustainability	• Lack of conscious drive towards sustainability.
• One approach to sustainable agriculture.	• Refers to main stream farming practices carried out by majority of farmers in the research sites.
• Low cost input approach, high reliance on re-cycling of on-farm resources.	• Spread and adoption facilitated by Government agencies.
• Inclusion of biological nitrogen fixing crops like legumes	
• Green manure crops need to be of prime importance	

Integrated pest management is probably the oldest and most widely recognized Extension Service programme devoted to low-input agriculture. However, only recently have the "non-chemical" approaches—such as cultural, mechanical, and biological—within the IPM framework have been emphasized over the chemical component. Some programmes, in fact, are now termed "biologically-intensive IPM."

The two-fold objective of LEISA programme was to establish a form of agriculture that

1. Would be productive and profitable for current farmers and
2. Conserve resources for future generations.

Low input farming systems "seek to optimise the management and use of internal production inputs (i.e., on-farm resources) and to minimise the use of production inputs (*i.e.*, off-farm resources), such as purchased fertilisers and pesticides, wherever and whenever feasible and practicable to lower production costs to avoid pollution of surface and groundwater, to reduce pesticide residues in food, to reduce a farmer's overall risk, and to increase both short- and long term farm profitability" (Parr *et al.*, 1990). The intention of adopting this definition is to ultimately clarify some of the misconceptions of low-input agricultural systems. David Norman *et al.* (1997), for example, argue that the term "low input" is somewhat misleading and indeed unfortunate as it often implies that farmers should starve their crops, let the weeds choke them out, and let insects clean up what is left. In fact, the term low-input (as implied by the above definition) is used to refer to purchasing fewer off-farm inputs (usually fertilisers and pesticides), while increasing on-farm inputs (*i.e.*, manure, cover crops, and especially management).

Low-input farming

Farming for LIFE

According to less intensive farming and the environment in UK (LIFE) which conducted trials at Long

Ashton near Bristol highlighted that the results from three years of study, profitability can be maintained with less intensive production. Further, the herbicide use can be reduced by 19%, fungicide use by 84% and insecticide use by 100%. The practices of LIFE farming which are as follows.

- The use of only those chemicals which are more selective,
- The use of minimal dosage and frequency of application.
- Spot treatment or inter row applications
- Low volume sprays and more careful timing of application.

Besides, the following strategies are recommended in order to reduce inputs.

- Modified cropping sequences to increase crop diversity.
- Use of tillage systems that favour natural control of key animal pests, diseases and weeds, improve soil structure and reduce demand for external nitrogen;
- Development and use of pest thresholds with decision models and non- chemical methods to reduce agrochemical input;
- Modification of field margins to encourage natural enemies of pests.

Biological Husbandry

"Biological" farming has become synonymous with farmers using the "Reams fertility system" as the basis for crop production. Eco-agriculture is the term used to describe this system by the monthly Acres, U.S.A. The Reams system is based on the LaMotte-Morgan soil test and the use of rock phosphate, calcium carbonate, and compost to achieve nutrient ratios of 7:1 calcium to magnesium, 2:1 phosphorus to potassium, and so on. "Biological" farming allows the use of selected chemical fertilizers (avoiding disruptive materials such as anhydrous ammonia and potassium chloride) and adopts low-input approaches to use herbicides and insecticides.

Diagnostic instruments to monitor plant and soil conditions are frequently used in "Biological" farming; these include refractometers to monitor sugar content (Brix) in plant tissue sap; electrical conductivity meters to monitor ERGS (or energy released per gram of soil); ORPS meters (or oxygen reduction potential of soil); and radionics. Based on data gathered, foliar sprays containing biostimulants and soluble nutrients are applied. The Pandol Brothers, a large commercial fruit and vegetable operation in California, reduced their annual pesticide bill from $500,000 to $50,000 per year after adopting a "Biological" fertility programme.

Biological farming is a system of activating and harnessing of biological energy on a sustainable basis for the production of high quality and profitable agricultural produce. It is farming in harmony with nature. It is a holistic and sustainable farming system that improves soil health, which inturn improves plant, livestock and human health. The benefits of biological farming are numerous. To start with, it results in a reduced need for inorganic fertilizers and pesticides. Despite the reduced chemical inputs, biological farmers most often experience an increase in yields of higher quality. The consumer's demands for healthy food are growing at an exponential rate, for which they are ready to pay a premium price.

Important aspects of biological farming

1. Biology in the soil.
2. Mineralizing the soil
3. Soil structure improvement

Biological farming utilizes resources of both science and nature in a superior farming system. Biological farming is a process aimed at restoring or improving soil life, soil structure and mineralizing the soil. Biological farming is working with nature to maximize nature's power in producing a crop. Biological farmers avoid using harmful chemicals like herbicides, pesticides and certain fertilizes that destroy soil life. A biological farmer might use less poisonous chemicals in an emergency to save a crop, but will also take measures to restore soil life by using products like compost tea and microbes to "clean" the soil of harmful chemicals. Biological farmers understand that the need for harmful chemicals is greatly reduced by practicing natural soil fertility building practices.

Biodynamic Farming

Biodynamic is a science of life- force, a recognition of the basic principles at work in nature and an approach to agriculture which takes these principles into account to bring about balance and healing. In a very real way, biodynamic is an ongoing path of knowledge rather than an assemblage of methods and techniques.

Biodynamic farming was spawned by the late anthropologist, Rudolf Steiner and has grown and developed since 1922. The term *biodynamic* is taken from the Greek words *bios* meaning life and *dynamic* meaning energy. Hence biodynamic farming refers to "working with the energies which create and maintain life". The concept realization is that other energies contribute to a plant's growth. Because of the differences in these contributing energies, planting your crop one day will be totally different than planting it on another day.

Biodynamic farming should be experienced. Hence, one can best learn through observation and feedback from an already practicing biodynamic farmer. There is an integrated relationship between plant, animal and soil that must be understood.

It is a method of crop production which is based on sound principles of soil biotechnology and soil microbiology. It does not require sophisticated facilities and most of the manures and biopesticides are prepared on the farm itself. Biodynamic preparations are components of biodynamic farming are to restore the soil and the organic matter in the form of humus, increasing microbial population, skillful application of factors contributing to soil life and health, treating manure and compost in biodynamic way. The biodynamic farming can be defined as "systematic and synergistic way of harnessing energies from cosmos, mother earth, plants and cow" for quality production and proper maintenance of ecosystem.

Plant life is intimately bound up with the life of the soil

Biodynamics recognizes that soil itself can be alive and this vitality supports and affects the quality and health of the plants that grow in it. Therefore, one of the biodynamic fundamental efforts is to build up stable humus in our soil through composting.

In a nutshell, biodynamics can be understood as a combination of "biological and dynamic" agriculture practices. "Biological" practices include a series of well-known organic farming techniques that improve soil health. "Dynamic" practices are intended to influence biological as well as metaphysical aspects of the farm (such as increasing vital life force), or to adapt the farm to natural rhythms (such as planting seeds during certain lunar phases).

The following information gives a brief summary of biological and dynamic farming practices.

a) Biological practices

Green manures
Cover cropping
Composting
Companion planting
Integration of crops and livestock
Tillage and cultivation

b) Dynamic practices

Special compost preparations
Special foliar sprays
Planting by calendar
Peppering for pest control
Homeopathy
Radionics

History of biodynamic farming

Biodynamic agriculture is an advanced organic farming system that is gaining increased attention for its emphasis on food quality and soil health.

Biodynamic agriculture developed out of eight lectures on agriculture given in 1924 by Rudolf Steiner (1861-1925), an Austrian scientist and philosopher, to a group of farmers near Breslau (which was then in the eastern part of Germany and is now Wroclaw in Poland).

The Agriculture Course lectures were taught by Steiner in response to observations from farmers that soils were becoming depleted following the introduction of chemical fertilizers at the turn of the century. In addition to degraded soil conditions, farmers noticed a deterioration in the health and quality of crops and livestock. Thus, biodynamic agriculture was the first ecological farming system to develop as a grassroots alternative to chemical agriculture.

Steiner was convinced that the food in his society was degrading and he believed the source of the problem was artificial fertilizers and pesticides, however he did not believe this was because of chemical or biological properties relating to the substances involved, but for *spiritual* shortcomings in these substances. Steiner considered the world and everything living in it as primarily spiritual in nature, the physical and thus chemical or biological processes involved were secondary. He also believed that living matter was different from dead matter, a viewpoint commonly referred to as vitalism. Many of Steiner's writings describe energy flows radiated from the earth akin to the so-called Odic force.

Another aspect of the idea is that the farm as a whole is a living system and therefore should be closed self-nourishing system, which the preparations nourish. Disease of organisms is not to be tackled in isolation but is a symptom of problems in the whole organism.

Thus, the concept of agricultural biodynamic comes to play a key role not only in the feature of organic farming practices, but also in the conversion to more environmentally and socially conscious ways of cultivating crops. Biodynamic is the oldest non- chemical agricultural movement, compared to the organic agriculture movement. Just recently, Steiner's basic ecological principle of biodynamic is to conceive of the farm as an organism, a self- contained entity. Biodynamic parallels organic farming in many ways but is set apart from other organic agriculture systems by its association with the spiritual science of anthroposophy founded by Steiner. It is based on "biological" practices which are defined as physical, technical actions and "dynamic" practices regulated by the weather, the seasons and what Steiner calls the "metaphysical." "Biodynamic" focuses on the following ideas which should be implemented through specific farm practices: planting according to the cosmos and the seasons, crop rotation, companion planting, soil fertility as the basic of healthy plants, composting, life forces beyond those which are physical and healthy crops, the farm as the unit and the farmer as an individual in that unit which ultimately, focuses on the earth's ability to heal itself and to rejuvenate its nutrients with the help of holistic orthodox farming practices.

Principles

A basic ecological principle of biodynamics is to conceive the farm as an organism, a self-contained entity. A farm is said to have its own individuality. Emphasis is placed on the integration of crops and livestock, recycling of nutrients, maintenance of soil and its health and wellbeing of crops and animals; the farmer too is part of the whole. Thinking about the interactions within the farm ecosystem naturally leads to a series of holistic management practices that address the enviromental, social and financial aspects of the farm and the farmer.

A fundamental tenet of biodynamic agriculture is that the food raised biodynamically is nutritionally superior and tastes better than foods produced by conventional methods. This is a common thread in alternative agriculture, because other ecological farming systems make similar claims for their products. Demeter, a certification programme for biodynamically grown foods, was established as early in 1928. As such, Demeter was the first ecological label for organically produced foods.

Biodynamic farming involves many principles and among them Zodiac principles are of prime importance. Zodiac principles can be explained through following ways.

i) Harnessing energies from cosmos

The ultimate fine- tune of biodynamic principles lies in harnessing cosmic and earthy energies which influences the growth of a particular part of a plant. The earthy forces of moon, mercury and Venus soak into the earth from the air above and cosmic forces of mars, Jupiter and Saturn moves upward from the rocks below. They interact in the region of clay so that the plants grow out of it. The light of sun, moon, planets and stars reaches to the plants in regular rhythms. Each contributes to the life, growth and form of the plant. Since ancient time, they have been divided in to inner planets which works directly through atmosphere or indirectly through water, humus or calcium (limestone, potassium and sodium) on growth of plants. The influence of mars, Jupiter and Saturn are channeled through warmth and silica (quartz, feldspar, mica),they stream in through silica contents of soil and on plants being expressed in colors of flowers and in fruits and seed production. By understanding gesture and effect of each rhythm, agricultural activities like soil preparation, sowing, intercultural operations and harvesting need to be programmed accordingly.

Biodynamic calendar

Moon in zodiac constellation (cycle 276.3 days)

The zodiac is a belt of fixed stars, which are in groups and are known as constellations. This belt lies behind the elliptical path of sun, the planets move in front of these constellations. Because of the different size of the constellation, the moon is in front of a constellation for a shorter or a longer duration, which may vary from 1.5 to 3.5 days. As the moon is in front, these fixed stars generate certain favorable effect on the plants. The system has 12 constellations, each of which have some qualities and they are related to four elements *i.e* earth, water, fire and Air. These four elements can be placed in relation to 4 parts of the plant.

Table. Interaction Element and Constellation on Plant Part

Elements	Plant parts	Functions	Constellations
Earth	Root	Control root functions, extraction of nutrients development of different plant parts	Taurus, Virgo, Capricorn
Air /light	Flower	Respiration	Gemini, Libra, Aquarius
Water	Leaf	Transportation of nutrients	Cancer, Scorpio, Pisces
Fire	Fruit/ Seed	Photosynthesis and metabolism	Aries, Leo, Sagittarius

ii) Harnessing energies from plant

Plants when grown on soil take nutrients and water from earth and perform the process of photosynthesis utilising sun's energy and completing its life cycle. But at the end, their vegetative parts get recycled in the soil sooner or later in different forms. The animals manure and other animals wastes are usually converted into a stable humus through fermentation process. Composting the organic materials will avoid nutrient losses oxidation or if these are applied directly on the ground. Appropriate crop rotation, inclusion of legumes as green manure, inter crops, cover crops, crop residues provide humus, which are essential for proper activity of microorganisms in the soil.

iii) Harnessing energies from cow

In biodynamic system cow horn, cow dung and cow urine are being used for preparation of a number of products to provide nutrition to the soil and control of pests and diseases. Cow dung and cow urine are the component of cow pat pit (CPP), BD liquid manures and pesticides. Their brief account is summarised as follows.

Role of cow dung

- Cow dung is basic component in cow pat pit and BD-500. Both preparations are used to improve the biological properties of the soil.
- Cow dung cake and ghee are basic components of Agnihotra in homofarming.
- In Japan, cow dung is used to get protection from atomic emissions.
- Cow dung is the best soil conditioner.
- CISH identified new actinomycetes, which are helpful in controlling many plant gummosis, dieback) diseases.

Importance of cow urine

- Besides, copper, cow urine contains iron, calcium, phosphorus, carbonic acid potash and lactase.
- It contains 24 types of salts and medicines made from cow urine are used to cure several diseases.
- Cow urine is disinfectant and it can be used as a prophylactic measure to purify and improve the soil fertility.
- Addition of cow urine in compost pit hastens the composting process (period reduced by 15 –20 days) and helps in further multiplication of worms.

In biodynamic practice cow dung and horn are utilized for the preparation of BD 500, BD 501 and cow pat pit. These are the basic preparation and are helpful in harnessing earthy and cosmic energies. Brief accounts of these are enumerated below.

Biodynamic preparations

The biodynamic preparations

A distinguishing feature of biodynamic farming is the use of nine biodynamic preparations described by Steiner for the purpose of enhancing soil quality and stimulating plant life. They consist of mineral, plant, or animal manure extracts, usually fermented & applied in small proportions to compost, manures, the soil, or directly onto plants, after dilution and stirring procedures called dynamizations (BDAI, Training Manual).

Fertilizers

Steiner prescribed eight different preparations for fertilizers which were allowed for use in biodynamic agriculture and gave great details of how these were to be prepared. The substances are numbered 500 through 507, whereof the first two are used for preparing fields whereas the latter six are used for making compost:

Field preparations

Field preparations, for stimulating humus formation

- 500: (horn-manure) a humus mixture prepared by stuffing cow manure into the hallow portion of a cow horn and then burried into the ground (40-60 cm below the surface) in the autumn and allowed for decomposition during a winter.
- 501: Crushed powdered quartz that can be mixed with 500 but usually prepared on its own (mixture of 1 tablespoon of quartz powder to 250 litres of water). The mixture is sprayed under very low pressure over the crop during the wet season to prevent fungal diseases. It should be sprayed on an overcast day to prevent burning of the leaves.

Both 500 and 501 are used on fields by stirring the contents of a horn in 40-60 litres of water for an hour and whirl it in different directions every second minute. About 4 horns are used for each hectare of soil.

Compost preparations

Compost preparations, used for preparing compost

- 502: Yarrow blossoms (*Achillea millefolium*) are stuffed into urinary bladders from Cervus elaphus, Red Deers, placed in the sun during summer, buried in earth during winter and retrieved in the spring.
- 503: Chamomile blossoms (*Chamomilla officinalis*) are stuffed into small intestines from cattle buried in humus-rich earth in the autumn and retrieved in the spring.
- 504: Stinging nettle (*Urtica dioca* and the whole plant in full bloom) is stuffed together under ground surrounded on all sides by peat for a year.
- 505: Oak bark (*Quercus robur*) is chopped in small pieces, placed inside the skull of some domesticated animal, surrounded by peat and buried in earth in a place where lots of rain water runs by.
- 506: Dandelion flowers (*Taraxacum officinale*) is stuffed into peritoneum from some cattle is buried in earth during winter and retrieved in the spring.
- 507: Valerian flowers (*Valeriana officinalis*) is extracted into water.

One to three grams (a teaspoon) of each preparation is added to a dung heap by digging 50 cm deep holes with a distance of 2 meters from each other, except for the 507 preparation, which is stirred into 5 litres of water and sprayed over the entire compost surface. All preparations are thus used in homeopathic quantities and the only intent is to strengthen the life forces of the farm, i.e. the preparations fulfill spiritual goals and nothing else.

Dealing with pests and weeds

Biodynamic agriculture use techniques reminiscent of the fertilization for pest control and weed control, most of these techniques include using the ashes of a pest or weed that has been trapped or picked from the fields and ceremonially burnt. Steiner sees pests and weeds as a result of imbalance between life forces emanated from the earth.

Since Steiner viewed the full moon, Venus and Mercury as cosmic powers influencing the fertility of plants, the biodynamic techniques for pest control involves blocking the fertility influence from said planets on different pests. Steiner dictates that this is achieved in different ways for pests and weeds:

- Weeds are combated by collecting seeds from the weeds and burning them above a wooden flame. The ashes from the seeds is then spread on the fields, which will according to biodynamic philosophy block the influence from the full moon on the particular weed and make it infertile.
- Pests such as insects or Apodemus (field mice) have more complex processes associated with them depending on what pest is to be targeted. For example field mice are to be countered by deploying ashes prepared from field mice skin when Venus is in the scorpius.

There are a series of preparations made from various medicinal herbs, in such a way as to enhance their inherent availability. These herbs are associated with constellations and influence availability of macro and microelements. Incorporation of these preparations catalyses the fermentation process. Biodynamic preparations, their association with constellation and parts used.

The details of biodynamic preparation are as follows and their significance are explained as follows.

Cow horn manure (BD-500)

It is basically fermented cow dung. It is usually the first preparation used during the change over to organic/

biodynamic system, This is fundamental biodynamic field spray preparation. Specially prepared manure is made into a spray to spray to vitalize the soil to enhance seed germination, root formation & primary root development. If possible it should be sprayed four times in a year. The best times are in autumn (October) and again in the spring (February & march) for spraying, 25 g BD 500 is dissolved in 13.5 liters of water is plastic buckets by making vortex in clock and anticlockwise direction from one hour in the evening. The basic principle of stirring in clock and anticlockwise direction is that while reverse process chaos is created for a moment . During this process cosmic forces are absorbed and water becomes active and preparation gets oxygenated. Stirring small qualities of material in large quantity of water is called "dynamization". This process transfers the forces of energy from the preparation to the water itself. Spraying of BD- 500 is done at the time of field preparation in evening during descending period of the moon.

Steps in preparation

- Cows horns are cleaned properly with water , precaution is to be taken while collecting the horn remains solid from proximal end and their rings are at distal end.
- Cleaned cow horn are filled with fresh cow dung (especially from lactating and indigenous one) and buried at 30 cm depth in the soil in root free zone in descending period of moon during the month of March and April.
- After 6 months of incubation, horns should be left for some more periods and again is to be taken out during descending period of moon.
- If decomposition of dung is not proper, cow dung should be left for some more periods and again is to be taken out during descending period of moon.

- Properly decomposed compost is to be stored at cool place in eastern pot.

Thimmaiah (2001) observed the microbial activity of BD -500 during stirring and very interesting response has been obtained . It is interesting to observe that during stirring period, there was a corresponding increases in the number of cfu's of bacteria, actinomycetes and fungi in one hour of stirring.

Cow dung silica (BD 501)

In this finely mountain quartz crystal (silica) after proper incubation is made into spray to benefit plants. Its action is to strengthen the effect of light and warmth on the plants and promotes healthy growth. Its helps in improvement in protein and sugar (brix) level, metabolism mechanical rigidity, growth, tolerance to environmental stresses (frost, drought, salinity, mineral toxicity and deficiency) and resistance to fungal attack. It also improves the taste, color and aroma and shelf life of produce.

Steps in preparation

- After taking out of cow horn manure (BD 500), cow horns are thoroughly cleaned with water.
- Cow horns are filled with silica powder paste and buried in same pit where cow horns were buried for the preparation of BD 500 during ascending period of moon in March and April.
- After 6 months of incubation, horns are taken out in Oct-Nov, during the ascending period of moon.
- Light yellowish silica powder is taken out from the horn and stored in light near the house wind in glass jar.
- The horn silica BD 501 catalyses cosmic forces of the outer planets of Mars, Jupiter and Saturn and applied in the Morning, particularly in summer when forces from within the earth well forth in the atmosphere.

BD 501: works on the photosynthesis process in the leaf. It strengthens the quality of plant product and encourages the development of fruit and seeds For maximum effect, the BD 501 should be applied on the leaves in the form of 'mist' in the morning at the sunrise and the best constellation in moon in opposite to Saturn.

BD 502: It is made from the flower of yarrow (*Achillea millefolium*), which helps in process of K, S and trace elements. During spring or early summer, harvest and naturally cry enough wild white yarrow florets to stuff in a dried inflated deer, stage or moose bladder. Before stuffing, soften the dried with Luke warm rainwater. Moisten the dried yarrow florets with tea made of yellow florets rainwater. Discard the florets used to make the compost. Press the moistured florets together before stuffing them into the bladder. When stuffed to the point of stretching, sew it closed with thread. Hang the stuffed bladder six feet above the ground in full sun all summer. In the fall of summer, bury it six to twelve inches deep in fertile soil. A year after first stiffed into the bladder, remove the yarrow and store in an earthern pot.

BD 503: It is made from the flowers of German chamomile. Which helps in process of Ca and N. Collect and dry naturally wild Chamomile flowers as early in the year as possible during spring season. The flowers are dried in shade and airy place. Stuff the moistened flowers into the intestine of a recently died cow. Bury this "Sausage" through the winter in fertile soil. In the spring take out and store the compost in an earthen pot until it is needed for use.

BD 504: It is made from stinging nettle, which brings process of calcium. Collect whole of the mature stinging nettle, before flowering. Allow the harvest to fade slightly then, just before winter, bury them for a year in rich soil. The fried nettle is stuffed into a rough unglazed earthen pot and buried infertile soil in autumn. It is sieved and stored. it contains sulphur, iron, potash and it is influenced by Mars planet. It is used as a haemostatic and improves milk flow for nursing.

BD 505: It is made from the bark of Himalayan Oak bark into the skull of any domestic animal such as cow, bull, goat, sheep. All the flesh and brain material is cleared before use. The stuffed skull needs to be shallowly buried where water flows. For lack of an optimal burial site, use a barrel with decaying plant substances and exposed to rain. The idea is bury the skull in an environment of Slimy rotting mulch, In the spring the result is humus high in microorganisms and calcium , which enhances the compost pile. Store in an earthen pot until required.. It contains mainly calcium of about 77%, it has an antifungal action and it is connected with moon forces and lifts the pH of a soil.

BD 506: It is made from the flowers heads of Dandelion. It is specialist in silica acid and potassium. It is used medicinally for liver diseases. It attracts the forces of outer planets like Jupiter.

BD 507: It is made from the flowers of valerian plant; this is connected at Saturn planet. Valerian is commonly found in Kashmir early on spring morning, harvest the florets grind the flowers and dilute the water at 1:4 ratio. This can be done at any time and restored in earthern pot, stir 10 ml into 30 liters of water and sprinkle over the compost.

BD 508: It is prepared from horsetail or casuarina leaves can also be used as a tea to control fungus in early season. Collect and dry green shoots of horse tail. It contains upto 70% silica, traces of alkaloids, tannins, manganese, potassium, sulphur and magnesium. Incorporating 100 g of this in five liter of water and slowly boils it. Stir BD 508 from 10 minutes just before application. Sprinkle or spray on soil around plants on full moon day.

Biodynamic compost

Biodynamic compost is a fundamental component of the biodynamic method; it serves as a way to recycle animal manures and organic wastes, stabilize nitrogen and build soil humus and enhance soil health. Biodynamic compost is unique because it is made with BD preparations 502-507. Together, the BD preparations and BD compost may be considered the cornerstone of biodynamics. Here again, "biological" and "dynamic" qualities are complementary: biodynamic compost serves as a source of humus in managing soil health and biodynamic compost emanates energetic frequencies to vitalize the farm.

It is enriched compost, act as a soil conditioner. BD compost is prepared by using green (nitrogenous material) and dry leaves (carbonaceous material) Integrating with cow dung slurry and BD 502-507 catalyses the decomposition process.

Steps in preparation

- Place 5 m long thick roof on higher elevation.
- Spread 20 cm thick layer of dry grass on wood in the size of 5 x 2.5 m.
- Sprinkle 100–150 liters of water mixed with dung on the grasses.
- Again 20 cm thick layer of green grasses are spread equally on the heap and sprinkle 100–150 liters of water with dung.
- Above process is repeated to the height of 1.5 m.
- Incorporate BD set (502 - 507) in the heap and plaster with mixtures of dung and soil.

Cow pat pit (CPP)

It is biodynamic field preparation, a good soil conditioner. This is mainly used for seed treatment and foliar application and prepared throughout the year.

Steps in preparation of CPP

- Preparation of a pit of size 60 x 90 x 45 cm in shade and root free zone.
- Pasting of inner wall with fresh cow dung.
- It is prepared with 60 kg of cow dung mixed thoroughly with 250 gm each of bentonite and eggshell powder and filled in the pit.
- Two sets of BD preparation (502 and 506) are released in the mixture by making hole.
- BD 507 @ 10 ml dissolved in 2-3 liters of water mixing thoroughly in clock anti clock wise from 10–15 minutes and sprinkled.
- Mixture is covered with gunny sac.
- Compost gets ready in 75–90 days depending upon the temperature.
- It is used by dissolving one kg CPP in 40-45 liters of water over night and sprinkled in the next morning as field spray on the plants.

Bio-dynamic tree paste (Preparation)

- Sand bentonite powder and fresh cow dung are taken 1:1:1 ratio and thoroughly mixed .
- 25 gm BD 500 stirred from one hour in 13.5 of water poured in the mixture.
- Paste is prepared by mixing of required quantity of water.
- Small amount of paste is taken in the plastic bucket diluted with water to thick solution and pasted on the truck with help of brush.

Uses

- It nourished strengths and protects the bark and cambium of tree to make it healthy.

- Seals and heals wounds.
- Prevention and control of diseases.
- On application after pruning stimulates tree growth.

Biodynamic liquid manure and pesticides

Liquid manures are prepared using a different material that is leaves of leguminous tree, neem, leaves, fish wastes, castor leaves and other medicinal plant parts.

- A plastic drum with capacity of 200 liters is taken.
- Leaves are chopped off in pieces and put in the drum.
- Add 5 liters of cow urine and 2 kg of cow dung, drum is filled up with 125 liters of water.
- BD–502 to 506 are hanged in the drum like tea bags or put inside in the leaves packets.
- BD 507 is mixed in 2–3 liters of water for 15–20 minutes and poured in the drum .
- Drum is covered with gunny sac.
- Liquid manure gets in 8–12 weeks.
- One liter liquid manure is dissolved in 4–5 liters of water and sprayed on the plant.

Ramakrishnappa (2000) conducted exhaustic studies on the effect of biodynamic sustainable organic farming system in cultivation of potato, cabbage, carrot, fresh bean, rosemary, thyme, carnation and gerbera and studied extensively at research stations and in farmers field. The experiments consisted of the following four treatments *viz.,* (i) biodynamic organic farming, (ii) conventional farming (iii) biodynamic + conventional and (iv) absolute control. The results of the experiments revealed that biodynamic sustainable organic farming system increased the cumulative yield of potato by 16 per cent under irrigated and 17.9 per cent under rainfed conditions. The practice of biodynamic organic farming system in potato cultivation has recorded the net profit of Rs.1,10,000 under irrigated condition and Rs.60,000 under rainfed conditions, whereas the conventional treatment recorded Rs,45000 and Rs,16500 as net profit under irrigated and rainfed condition, respectively. This system of cultivation was also found to increase contents of carbohydrates (60%), protein (66%), vitamin C (74%) and B-carotene (82%) over the conventional system in carrot crop. In cabbage, 14.3 per cent and 28.3 per cent of cumulative organic farming recorded in biodynamic organic farming system under irrigated and rainfed systems, respectively. In French beans, cumulative yield increase was 34.8 per cent and 26.0 per cent in biodynamic organic farming system under irrigated and rained system respectively. The biodynamic organic farming system was also experimented in cut flower production. In carnation this system recorded significantly higher number of flowers per plant (17.8) with increased shelf-life of flowers. The same trend followed in gerbera also. The organic practices increased the stalk length of flowers (47.8 cm) and number of flowers (32.3/plant), which were highly significant than other treatment. The biodynamic organic treatment has recorded 32 per cent higher flower yield than the conventional treatment. the shelf-life of the gerbera flowers obtained with the organic treatment was 14 days which was 6 days more than conventional system. In general, the quality, colour and appearance of the flowers produced through organic methods in carnation and gerbera were excellent. In medicinal plants, the biodynamic organic farming system increased the herbage yield by 18.5 per cent in rosemary and oil content by 1.1 per cent which is 33 per cent higher than the conventional.

Table. Nutrient status of BD preparations and composts

Preparations	N (%)	P (%)	K (%)	Zn (%)	Cu (%)
FYM	0.5	0.35	0.50		
FYM fortified with CPP	1.02	0.38	0.32	0.05	0.009
Vermicompost	0.92	0.39	0.71	0.04	0.09
BD liquid manure (neem)	1.78	0.62	0.78	2.00	0.31
BD liquid manure (subabul)	1.93	0.87	0.62	1.8	0.31
BD liquid manure (lantana)	1.00	0.32	0.27	2.12	0.17

Ramkrishnappa (2000)

Integration of biofertilizer and biocontrol agents and the biodynamic organic farming systems enhanced the soil microbial activity (indicated by two to three- fold increase in microbial population *viz.*, bacteria, fungi, actinomycetes, diazotrophs and phosphate solubilizing microorganisms) to favour the plant vigour, plant growth and enhance the yield and quality parameters of all the crops.

Biodynamic farming involves restoring the soil a balanced living condition through the application and use of the completely digested form of crude organic matter known as stabilized humus. Crop rotation, correct composting and proper intercropping can all contribute to a healthier biodynamic yield.

Organic Farming

Organic farming is a form of agriculture that relies on ecosystem management and attempts to reduce or eliminate external agricultural inputs, especially synthetic ones. It is a holistic production management system that promotes and enhances agro-ecosystem health, including biodiversity, biological cycles and soil biological activity.

In preference to the use of off-farm inputs, organic farming emphasizes management practices, taking into account the regional conditions and locally adapted systems. Utilizing both traditional and scientific knowledge, organic agricultural systems rely on agronomic, biological, and mechanical methods (these may require external inputs of nonrenewable resources, like tractor fuel), as opposed to using synthetic materials, to fulfill any specific function within the system. Organic farming is also associated with support for principles beyond the cultural practices, such as fair trade and environmental stewardship, although this does not apply to all organic farms and farmers.

According to a study on "The World of Organic Agriculture - Statistics and Emerging Trends 2005", currently more than 26 million hectares of farm land are under organic management worldwide. This is two million hectares higher than the reported once during 2004 accounting for an increase of almost ten per cent.

In terms of organic land, Australia leads the pack with 11.3 million hectares, followed by Argentina (2.8 million hectares) and Italy with more than one million hectares. In Asia, China, India and Japan are the largest organic producers. Japan is the region's largest market and buys majority of Asia's production.

Definition of organic agriculture as specified by the national programme for organic production

"It is a system of farm design and management to create an ecosystem which can achieve sustainable productivity without the use of artificial external inputs such as chemical fertilizers and pesticides".

Organic farming is the most widely recognized alternative farming system. Modern organic farming evolved as an alternative to chemical agriculture in the 1940s, largely in response to the publications of J.I. Rodale in the U.S., Lady Eve Balfour in England, and Sir Albert Howard in India.

The definition of organic farming is very complex and many scientists have defined in their own way which are mentioned as under.

Lampkin (1990) defines organic farming as a production system which avoids or largely excludes the use of synthetic compounded fertilizers, pesticides, growth regulators and livestock feed additives. To the maximum extent possible, organic farming concentrates on crop rotations, residues, manures, legumes, green leaf manures, green manuring off-farm organic wastes and biological methods of pest control. The prime objective is to maintain soil productivity and tilth, to supply plant nutrients and to control insects, weeds and other pests.

Organic farming does not totally exclude the elements of modern agriculture. According to Fantilanan (1990) organic farming is a matter of giving back to nature by the way taken from it. It is inexpensive, profitable and sensible. Organic farming is not mere nonchemicalism in agriculture, but is a system of farming based on integral relationship. Therefore, one should know the relationships of soil, plant, water and microflora and the overall relationship between plants and the animal kingdom of which man is the apex animal. It is the totality of these relationships which is the backbone of organic farming.

The principle elements, objectives, salient characteristics, components and advantages are briefly explained as under.

Main principles of organic farming

The main principles of organic farming are as follows.

- Maintaining a living soil

- Making available all the essential nutrients
- Organic mulching for conservation
- Attaining sustainable high yield

Objectives

The objective of organic agriculture has been expressed in the standard document of the International Federation of Organic Agriculture Movement (IFOAM) as follows.

- To produce food of high nutritional quality in sufficient quantity.
- To work with natural systems rather than seeking to dominate them.
- To encourage and enhance the biological cycles within farming system involving microorganism, soil flora and fauna, plants and animals.
- To maintain and increase the long term fertility of soils.
- To use, as far as possible, renewable resources in locality organized agricultural systems;
- To work, as much as possible, within a closed system with regard to organic matter and nutrient elements.
- To give all livestock, conditions of life that allow them to perform all aspects of their innate behaviour.
- To avoid all forms of pollution that may result from agriculture techniques
- To maintain the genetic diversity of the agricultural system and its surroundings including the protection of plant and wildlife habitats.
- To allow agricultural producers an adequate return and satisfaction from their work including a safe working environment and
- To consider the wider social and ecological impact of the farming system.

Essential characteristics of organic farming

- Maximal but sustainable use of local resources.
- Minimal use of purchased inputs, only as complementary to local resources.
- Ensuring the basis of biological functions of soil, water, nutrients, human continuum.
- Maintaining a diversity of plant and animal species as a basis for ecological balance and economic stability.
- Creating an attractive overall landscape which gives satisfaction to the local people.
- Increasing crop and animal intensity in the form of polyculture, agroforestry system, integrated crop/ livestock systems etc. to minimize risks.

Components of organic farming systems

- Crop and soil management
- On farm waste recycling
- Non chemical weed management
- Biological pest control

Advantages of organic farming

- Organic manures produce optimal conditions in the soil for high yields and good quality crops.
- They supply all the nutrients required by the plant (NPK, secondary and micronutrients).
- They improve plant growth and physiological activities of plants.
- They improve the soil physical properties such as granulation and good tilth, giving good aeration, easy root penetration and improved water holding capacity.

- They improve the soil chemical properties such as supply and retention of soil nutrients and promote favourable chemical reactions.
- They reduce the need for purchased inputs.
- Organically grown crops are believed to provide more healthy and nutritionally superior food for man and animals than those grown with commercial fertilizers.
- Organically grown plants are more resistant to disease and insects and hence only a few chemical sprays or other protective treatments are required.
- There is an increasing consumer demand for agricultural produces which are free of toxic chemical residues. In developed countries, consumers are willing to pay more for organic foods.
- Organic farming helps to prevent environmental degradation and can be used to regenerate degraded areas.

Methods of organic farming vary. However, organic approaches share common goals and practices. In addition to the exclusion of synthetic agrochemicals, these include protection of the soil (from erosion, nutrient depletion, structural breakdown), promotion of biodiversity (e.g. growing a variety of crops rather than a single crop), and outdoor grazing for livestock and poultry. Within this framework, individual farmers develop their own organic production systems, determined by factors such as climate, market conditions, and local agricultural regulations.

It is important to make the distinction between organic farming and organic food. Farming is concerned with producing fresh products such as vegetables, fruits, meat, dairy, eggs for immediate consumption, or for use as ingredients in processed food. It is also important to note that organic farming is a reaction against the large-scale, chemical-based farming practices that have become the norm in food production over the last 80 years. The differences between organic farming and modern conventional farming account for most of the controversy and claims surrounding organic agriculture and organic food. Until recently, the comparison looked something like this:

Particulars	Organic	Conventional
Size	Relatively small-scale, independent operations (e.g. the family farm)	Large-scale, often owned by or economically tied to major food corporations
Methods	No use of purchased fertilizers and other inputs; low mechanization of the growing and harvesting process	Intensive chemical programs and reliance on mechanized production, using specialized equipment and facilities
Markets	Often local, direct to consumer, through on-farm stands and farmers' markets (see also local food), and through specialty wholesalers and retailers (eg: health food stores)	Wholesale, with products distributed across large areas (average supermarket produce travels hundreds to thousands of miles) and sold through high-volume outlets

Organic *v/s* natural farming

Organic farming is synonymous to natural farming. However, to make explicitly clear the differences, the following distinguishing features may be noted.

Standards and legislation relating to organic agriculture

Increasingly, organic farming is defined by formal standards regulating production methods, and in some cases, final output. Two types of standard exist, voluntary and legislated. As early as the 1970s, private associations created standards, against which organic producers could voluntarily have themselves certified. In the 1980s, governments began to produce organic production guidelines. Beginning in the 1990s, a trend toward legislation of standards began, most notably with the European Union.

Organic farming	Natural farming
Close to natural farming but does not have the philosophical overtone of natural farming	It is not alternative system of farming but part of philosophy of life involving continuous search to know the true spirit and form of nature.
Organic farming is not totally exclusive of modern farming. It involves tillage, weed control practices and use of chemicals.	It is devoid of all components of modern farming
It includes a soil building programme more intensive style of natural farming. Application of natural plant protection chemicals (which are not inorganic derivations), use of organic manures are permitted	It indicates a "do nothing' approach
Principle elements to be considered a) Maintaining a living soil b) Making available all essential nutrients c) Organic mulching	The essential principles are; 1. No cultivation 2. No chemicals 3. No weeding 4. No plant protection

In the 70s and 80s, organic certification of farms emerged as a marketing tool to insure foods produced organically met specified standards of production. The Organic Foods Production Act, included in the 1990 Farm Bill, enabled USDA to develop a national program of universal standards, certification accreditation, and food labeling. Implementation, initially scheduled for October of 1993, was delayed due to lack of funding and complexity of issues and is anticipated to take effect in 1995.

An international framework for organic farming is provided by the International Federation of Organic Agriculture Movements (IFOAM), the international democratic umbrella organization established in 1972. For IFOAM members, organic agriculture is based upon the Principles of Organic Agriculture and the IFOAM Norms. The IFOAM Norms consist of the IFOAM Basic Standards and IFOAM Accreditation Criteria.

Table. Some facts on International standards

1. IFOAM	• Established in 1972 • Headquarter in Germany • Umbrella organisation for organic Agriculture Association Developed international basic standards of organic agriculture • Established IFOAM accreditation programme (1992) to accredit certifying bodies. • Set up International Organic Accreditation Service (IOAS) in July 2001
2. CODEX	• Codex Alimentarious Commission – a joint FAO/ WHO • Intergovernment body • Established in 1962 • Produced a set of guidelines for organic production
3. EU regulation	• Laid out a basic regulation for European Union's organic standards in Council regulation NO. 2092/91 (June 1991). • Regulations give guidelines for the production of organic crops in the European community.
4. Demter	• Demeter International is a world wide network of 19 International certification bodies in Africa, Austalia, Europe. • Developed guideline for biodynamic preparation.
5. JAS	• A set of guidelines 'Japan Agricultural Standards' for organic production.

Source: Helga Muller, 2000

The codex guidelines for organic production system

- Enhance biological diversity within the whole system
- Increase soil biological activity
- Maintain long term soil fertility
- Recycle wastes of plant and animal origin inorder to return nutrients to the land, thus minimizing the use of non renewable resources.
- Rely on renewable resources in locally organized agricultrual systems.
- Promote the healthy use of soil, water and air as well as minimize all forms of pollution thereto that may result from agricultural practices.
- Handle agricultural products with emphasis on careful processing methods in order to maintain the organic integrity and vital qualities of the product at all stages.
- Become established on any existing farm through a period of conversion, the appropriate length of which is determined by site specific factors such as the history of the land, and type of crops and livestock to be produced.

The IFOAM Basic Standards are a set of "standards for standards." They are established through a democratic and international process and reflect the current state of the art for organic production and processing. They are best seen as a work in progress to lead the continued development of organic practices worldwide. They provide a framework for national and regional standard-setting and certification bodies to develop detailed certification standards that are responsive to local conditions.

Legislated standards are established at the national level, and vary from country to country. In recent years, many countries have legislated organic production, including the EU nations (1990s), Japan (2001), and the

US (2002). Non-governmental national and international associations also have their own production standards. In countries where production is regulated, these agencies must be accredited by the government.

India's National Organic Programme was developed and implemented by the Government of India through its Ministry of Commerce. The Ministry of Commerce established a National Steering Committee for Organic Production (NSCOP), whose members were drawn from the Ministry of Agriculture, Commodity Boards, Food Processing Industries, Forests and Environment, Science and Technology, Rural Development and Commerce, and Trade and Exports. In March 2000, the National Steering Committee laid down the National Programme for Organic Production (NPOP). The Indian NPOP was modelled after the IFOAM (International Federation of Organic Agriculture Movement).

In June 2001, the first export regulations for certified organic products were prepared and approved by the National Steering Committee. Through the National Programme for Organic Production, the NSCOP formulated a National Accreditation Policy and Programme and developed standards for organic production and processes as well as the regulations for use of the National Organic Certification Mark.

Government regulations regarding organic food production have been targeted towards the export market. To regulate the export of certified organic products, the Director General of Foreign Trade, Government of India, issued a Public Notice which state that "an agricultural product will be allowed to be exported as "organic product" only if it is produced, processed and packed under a valid organic certificate issued by a certifying agency duly accredited by the NSCOP". The NSCOP consisted of representatives from the Ministry of Agriculture,

Commodity Boards, Food Processing Industries, Forests and Environment, Science and Technology, Rural Development and Commerce, and Trade and Exports. The main objectives of the NPOP have centralized around the export market.

The aims of the National Program for Organic Production include:

- providing the means to evaluate certification programmes for organic agriculture and products as per the approved criteria;
- developing policies for the certification and development of organic products;
- producing the National Standards for Organic Products (NSOP);
- formulating the National Accreditation Policy and Programme (NAPP);
- accrediting certification programmes to be operated by Inspection and
- Certification Agencies;
- facilitating certification of organic products in conformity to the NSOPs;
- developing regulations for the use of the National Organic Certification Mark;
- encouraging the development of organic farming and organic processing.

In 2002, the United States Department of Agriculture (USDA) established production standards, under the National Organic Program (NOP), which regulate the commercial use of the term organic. Farmers and food processors must comply with the NOP in order to use the word.

Inspection and certification of organic products marketed domestically

The National Program for Organic Production (NPOP) was developed and implemented by the National Steering Committee for Organic Products (NSCOP), through the Government of India's Ministry of Commerce. It was the job of the NSCOP to formulate a National Accreditation Policy and Programme and to draw up the National Standards for Organic Products. The standards for the NPOP and the National Accreditation Policy were prepared on the basis of the guidelines evolved by the International Federation of Organic Agriculture (IFOAM), the EU regulations and the Codex Alimentarius Commission.

The National Accreditation Policy was approved on 25 May 2001 by the National Accreditation Body. The work of the National Accreditation Body includes;

- Drawing up procedures for evaluation and accreditation of certification programmes;
- Formulating procedures for evaluation of the agencies implementing the programmes;
- Accreditation of inspection and certification agencies.
- The National Accreditation Body has designated six Accreditation Agencies including;

  a. Agricultural Processed Food Products Export Development Authority (APEDA);

  b. Coffee Board;

  c. Spices Board;

  d. Tea Board;

  e. Coconut Development Board;

  f. Directorate of Cashew and Cocoa Development.

Domestic accredited inspection and certification bodies

- Association for Promotion of Organic Farming (APOF)
- Indian Organic Certification Agency (INDOCERT) [Cochin, Kerala State]
- Indian Society for Certification of Organic Products (ISCOP) [Tamil Nadu]
- International Resources for Fairer Trade [Mumbai]
- LACON GMBH [Kerala State]
- SGS India Pvt. Ltd. [Gurgaon]

International accredited inspection and certification bodies

- Bioinspecta [Cochin, Kerala State]
- ECOCERT International [Aurangabad, Maharashtra State]
- IMO Control Private Limited [Bangalore]
- Naturland - Association for Organic Agriculture [Haryana]
- SKAL International [Bangalore]

As of 1st October, 2004, under the National Project on Organic Farming the government has established one National Centre on Organic Farming in Ghaziabad, along with six Regional Centres (RCOGs) in Bangalore, Bhubaneswar, Imphal, Hissar, Nagpur and Jabalpur.

Furthermore, there are associations of organic producers, inspection and certification bodies. For example, to improve extension work at the field level, Indocert, is initiating the Indian Organic Advisors Association. This association aims to provide technical advice for farmers and will function as a platform for advice, information dissemination and training in the field of organic agriculture.

Permaculture

Introduction

The word "permaculture" was coined by Bill Mollision, an Australian ecologist and one of his students, David Holmgren. The term is a contraction of "permanent agriculture" or "permanent culture."

In the mid 1970s, two Australians, Dr. Bill Mollison and David Holmgren, started to develop ideas that they hoped could be used to create stable agricultural systems. This was a result of their perception of a rapidly growing use of destructive, industrial-agricultural methods. They felt these methods were poisoning the land and water, reducing biodiversity and removing billions of tonnes of soil from previously fertile landscapes. A design approach called 'permaculture' was their response and was first made public with the publication of *Permaculture One* in 1978.

The term *permaculture* initially meant "permanent agriculture" but this was quickly expanded also to stand for "permanent culture" as it was seen that social aspects were an integral part of a truly sustainable system. Mollison and Holmgren are widely considered to be the co-originators of the modern permaculture concept.

Permaculture has developed from its origins in Australia into an international 'movement'. English permaculture teacher Patrick Whitefield, author of *The Earth Care Manual* and *Permaculture in a Nutshell*, suggests that there are now two strands of permaculture: a) Original and b) Design Permaculture. Original permaculture attempts to closely replicate nature by developing edible ecosystems which closely resemble their wild counterparts. Design permaculture takes the working connections at use in an ecosystem and uses this as its basis. The end result may not look as "natural" as a forest garden, but still has an underlying design based on ecological principles.

The term 'Permanent agriculture' was first coined by Franklin Hiram King in his classic book *Farmers of Forty Centuries*. Published in 1911. In this context, permanent agriculture is understood as agriculture that can be sustained indefinitely.

This definition was supported by Australian P.A. Yeomans (*Water for Every Farm*, 1973) who introduced an observation-based approach to land use in Australia in the 1940's, based partially on his understanding of geology. Yeoman introduced Keyline Design as a way of managing the water supply of a site.

The work of Howard T.Odum also influenced, Holmgren. Odum's work focused on system ecology, in particular the Maximum power principle, which examines the energy of a system and how natural systems tend to maximise the energy embodied in a system. For example, the total calorific value of woodland is very high with its multitude of plants and animals. It is an efficient converter of sunlight to biomass. A wheat field, on the other hand, has much less total energy and often requires a large energy input in terms of fertiliser.

Permaculture is about designing ecological human habitats and food production systems. It is a land use and community building movement which strives for the harmonious integration of human dwellings, microclimate, annual and perennial plants, animals, soils and water into stable, productive communities. The focus is not on these elements themselves, but rather on the relationships created among them by the way we place them in the landscape. This synergy is further enhanced by mimicking patterns found in the nature.

A central theme in permaculture is the design of ecological landscapes that produce food. Emphasis is placed on multi-use plants, cultural practices such as sheet mulching and trellising and the integration of animals to recycle nutrients and graze weeds.

However, permaculture entails much more than just food production. Energy-efficient buildings, waste water treatment, recycling and land stewardship in general are other important components of permaculture. More recently, permaculture has expanded its purview to include economic and social structures that support the evolution and development of more permanent communities, such as co-housing projects and eco-villages. As such, permaculture design concepts are applicable to urban as well as rural settings and are appropriate for single households as well as whole farms and villages.

Permaculture defined

From bill mollison

Permaculture is a design system for creating sustainable human environment.

From the permaculture drylands Institute, published in "The permaculture Activist" (Autumn 1989)

Permaculture: the use of ecology as the basis for designing integrated systems of food production, housing, appropriate technology and community development. Permaculture is a system for caring the interacting elements with the environment in mutually beneficial ways.

From Lee Barnes (former editor of Katuah Journal and permaculture connections), Waynesville, North Carolina

Permaculture (Permanent agriculture or culture) is a sustainable design system stressing the harmonious interrelationship of humans, plants, animals and the Earth.

According to Bill Mollison

Permaculture principles focus on thoughtful designs for small scale intensive systems which are lobour efficient and which use biological resources instead of fossil fuels. The core of permaculture is a design and the working relationships and connections between all things. Each component in a system performs multiple functions and each function is supported by many elements. Key to efficient design is observation and replication of natural ecosystems, where designers maximize diversity with polycultures, stress efficient energy planning for houses and settlement, using and accelerating natural plant succession.

From Michael Pilarski, founder of Friends of the trees, published in International Green Front Report (1988)

Permaculture is the design of land use systems that are sustainable and environmentally sound; the design of culturally appropriate systems which lead to social stability; a design system characterized by an integrated application of ecological principles in land use; an international movement for land use planning and design; an ethical system stressing positivism and cooperation.

From a Bay area permaculture group brochure, published in west coast permaculture News & Gossip and sustainable living newsletter (Fall 1995)

Permaculture is a practical concept which can be applied in the city, on the farm and in the wilderness. Its principles empower people to establish highly productive environments providing food, energy, shelter and other material and non-material needs, including finance.

In the broadest sense, permaculture refers to land use systems which promote stability in society, utilize resources in a sustainable way and preserve wildlife habitat and the genetic diversity of wild and domestic plants and animals. It is a synthesis of ecology and geography of observation and design. Permaculture involves ethics of earth care because the sustainable use of land cannot be separated from life-styles and philosophical issues.

Characteristics of permaculture (Pilarski, M., 1994)

- permaculture is one of the most holistic, integrated system analysis and design methodologies found in the world.
- permaculture can be applied to create productive ecosystems from the human- use standpoint or to help degraded ecosystems.
- permaculture validates traditional knowledge and experience and incorporates sustainable agriculture practices and land management techniques and strategies from around the world. Permaculture is a bridge between traditional cultures and emerging man made cultures.
- permaculture promotes organics agriculture which does not use pesticides to pollute the environment.
- permaculture aims to maximize symbiotic and synergistic relationships between site components.
- permaculture is urban planning as well as rural land design.
- permaculture design is site specific, client specific and culture specific.

The practical application of permaculture

Permaculture is not limited to plant and animal agriculture, but also includes community planning and development, use of appropriate technologies (coupled with an adjustment of life-style) and adoption of concepts and philosophies that are both earth-based and people-centered, such as bioregionalism. Many of the appropriate technologies advocated by permaculturists are well known. Among these are solar and wind power, composting toilets, solar green houses, energy efficient housing and solar food cooking and drying.

Due to the inherent sustainability of perennial cropping systems, permaculture places a heavy emphasis on tree crops. Systems that integrate annual and perennial crops- such as alley cropping and agroforestry- take advantages of "the edge effect," increases biological diversity and offer other characteristics missing in monoculture systems. Thus, multicropping systems that blend woody perennials and annuals hold promise as viable techniques for large- scale farming ecological methods of production. For any specific crop or farming system (*e.g.*, soil building practice, biological pest control, composting)are central to permaculture as well as to sustainable agriculture in general.

Since permaculture is not a production system, *per se*, but rather a land use and community planning philosophy, it is not limited to a specific method of production. Furthermore, as permaculture principles may be adapted to farms or villages worldwide, it is site specific and therefore amenable to locally adapted techniques of production.

As an example, standard organic farming and gardening techniques utilizing cover crops, green manures, crop rotation and mulches are emphasized in permaculture systems. However, there are many other options and technologies available to sustainable farmers working within a permacultural framework (*e.g.*, chisel plows, no-till implements, spading implements, compost turners, rotational grazing). The decision as to which "system" is to be employed is site-specific and management dependent .

The ethics of permaculture

Core values

a. Earthcare – recognising that the earth is the source of all life (and is possibly itself a living entity) and that we recognise and respect that the Earth is our valuable home and we are a part of the earth including non living things such as plants, animals, land, water and air.

b. Peoplecare – supporting and helping each other to develop healthy societies.
c. Fairshares (or placing limits to consumption) - ensuring that the Earth's limited resources are utilized in ways that are equitable and wise.

The principles of permaculture design

The principles of permaculture design may be broadly understood through a design known as O'BREDIM design methodology which emphasis in the following ways.

Observation, Boundaries, Resources, Evaluation, Design, Implementation, Maintenance.

O = Observation allows to first see how the site functions within itself, to gain an understanding of its initial relationships. Some people recommend a year long observation of a site before anything is planted. During this period all factors, such as ley of the land, natural flora, can be brought into the design. A year allows the site to be observed through all seasons.

B = Boundaries refer to physical ones as well as the neighbours

R = Resources include the people involved, funds as well as the crop planning and its expected returns.

E = Evaluation of the first three will then allow you to prepare for the next three. This is a careful phase of taking stock of what you have at hand to work with.

D = Design is always a creative and intensive process and you must stretch your ability to see possible future synergetic relationships.

I = Implementation is literally the ground- breaking part of the process when you carefully dig and shape the site.

M = Maintenance is then required to keep your site at a healthy optimum, making minor adjustments as necessary.

The principles of permaculture provide a set of universally applicably guideline which can be used in designing sustainable habit distilled from multiple disciplines- ecology, energy conservation, landscape design and environmental science – these principles are inherent in any permaculture design , in any climate and at any scale.

Permaculture zone

Permaculture zones classify three dimensional areas according to the amount of human attention needed to maintain the sustainable function of each zone.

Zone 0 - The house, or home centre. Here permaculture principles would be applied in terms of aiming to reduce energy and water needs, harnessing natural resources such as sunlight and generally creating a harmonious sustainable environment in which to live, work and relax in.

Zone 1 - Is the zone nearest to the house, the location for those elements in the system that require frequent attention, or that need to be visited often.

Zone 2 - The vegetable garden, large scale compost bins and maybe bee hives.

Zone 3 - Is the area where crops are grown, both for domestic and trading purposes which even include orchards. After establishment, care and maintenance requirements are fairly minimal providing mulches, etc. are used. Watering or weed control is once a week or so.

Zone 4 - Is semi-wild. Used for timber production from coppice managed woodland and the placement of aquaculture ponds.

Zone 5 - The wilderness. There is no human intervention here apart from the observation of natural ecosystems and cycles. Here is where we learn the most important lessons of the first permaculture principle of working with nature, not against it.

Regenerative agriculture

Regenerative agriculture became the preferred term of the Rodale Institute in the late 1970s and 80s under the direction of Robert Rodale. Regenerative agriculture builds on nature's own inherent capacity to cope with pests, enhance soil fertility, and increase productivity. It implies a continuing ability to re-create the resources that the system requires. In practice, regenerative agriculture uses low-input and organic farming systems as a framework to achieve these goals.

Advantages

- A method of growing food that relies on the internal resources that exist naturally, including the land, sun, air, rainfall, plants, animals and people.
- Regenerating the soil, returning the land to its natural state.
- Regenerating the health of human beings by reducing or eliminating chemical fertilizers, pesticides and herbicides.
- Regenerating the wildlands and local environment. Wildlife and beneficial birds return to help keep down insects and other pests. Our local watershed has less pollution in the streams and creeks which come off our land.

It helps to regenerate the communities by recycling organic waste that would otherwise be an economic or environmental hazards.

Vedic Agriculture

The word VEDAS comes from the VID, meaning "to know", and hence Veda means (knowledge) VEDA is treasure of knowledge & Vedic agriculture is a scientific branch of it. Vedic Agriculture practices are mentioned in Rig-Veda, Krishiparshara, and Manusmriti, Agnipurana. & Vriksha Ayurveda.

The concept of vedic Agriculture is based on the unique philosophy of "Harmony" with natural forces *i.e.*, Panchamahabutas.

Homa organic farming

There are two basic energy systems in the physical world: heat and sound. In performing *yagna,* these two energies, namely, the heat from *yagna's* fire and the sound of the chanting of the *Gayatri* and other *Vedic Mantras,* are blended together to achieve the desired physical, psychological and spiritual benefits (Thimmaiah, 2001).

Interaction of energy and matter from Vedic viewpoint

Panchamahabuthas or Five elements	Functions	Plant parts involved
Earth or Prithvi	Controls root functions, extraction of nutrients and development of different plant parts	Roots, stems
Water or Jala	Transport nutrients	Leaves (Veins and petioles)
Fire or Teja	Photosynthesis Metabolism	Seeds, Fruits
Air or VAYU	Respiration	Leaves
Space or AKASH	Makes space available for performing the above functions	Vacuoles

Rishi krishi

Rishi Krishi is a turning point to agriculture science. It is based on vedic literature and cosmic energy. The aim of Rishi Krishi technique is to keep the soil alive forever with the help of cosmic energy as it the only source of plant growth.

It has been observed that more than three lakh farmers (up to Dec. 2003) are following this method and are getting bumper crops, at different places with variety of soils as well as variety of crops ranging from small farmers having a piece of the land to the landlords having hundreds acres of land.

Rishi krishi a system of Agriculture practiced in Maharastra is using Amrit pani (prepared by mixing 20 kg cow dung, 0.125 kg butter, ½ kg honey, ¼ kg ghee) and kept over night to treat seeds and for spraying on field crops to maintain soil fertility and crop yield (Despande, 2004).

Rishi krishi aims to maintain soil fertility and crop productivity.

Chemical-Analytical Data

Whole rice grain		Rice ash	
Contents	%	Contents	%
Protein	6.94	Potash	21.37
Fat	0.5	Sodium	05.50
Starch	77.61	Calcium	03.24
Water	14.41	Magnesium	11.00
Saw dust	00.08	Ferric oxide	01.23
Others	00.46	H_3PO_4	53.68
		Silica	02.70
		Sulfuric	00.62
		Chlorine	00.01

Cow ghee		Agnihotra ash	
Contents	%	Contents	%
Butyric acid Capric acid Caproic acid	04.00 02.00 02.00	Nitrogen Phos.pentoxide Potash Others	00.34 97.00 2.32 00.34

Cow Dung	Agnihotra Gases
Menthol, phenol formalin, phosphoric acid, potash, ammonia, nitrogen	Formaldehyde, ethylene, oxide, propylene oxide, B-propiolatone

(Jayant Potdar, 1992).

Panchagavya

In Sanskrit, Panchagavya means the blend of five products obtained from cow (All these five products are individually called Gavya and collectively termed as Panchagavya). It contains ghee, milk, curd, cow dung and cow's urine. Panchagavya had reverence in the scripts of Vadas (divine scripts of Indian wisdom) and *Vrkshyurveda* (*Vrkshyurveda* means plants and ayurveda means health system). The texts on Vrkshyurveda are systematizations of the practices the farmers followed at field level, placed in a theoretical framework and it defined certain plant growth stimulants; among them Panchagavya was an important one that enchanced the biological efficiency of crop plants and the quality of fruits and vegetables (Natarajan, 2002).

Panchagavya spray was effective on all the the crops under evaluation than the foliar spray of recommended nutrients and growth regulators (RFS) for each and found advantageous since it recorded higher growth and productivity than no Panchagavya spray. Individually or system as a whole, biogas slurry with Panchagavya combination is adjudged as the best organic nutritional practice for the sustainability of maize, sunflower, green gram system by its overall performance on growth, productivity and quality of crops, the soil health and economic (Somasundaram, 2003).

Ingredients used in preparation of Panchagavya are

S. No.	Ingredients	Quality
1.	Fresh cow dung	7 kg
2.	Cow urine	3 liters
3.	Cow milk	2 liters
4.	Cow curd	2 liters
5.	Cow ghee	1kg
6.	Sugarcane juice	3 liters
7.	Tender coconut water	3 liters
8.	Ripe banana	12
9.	Toddy (if available)	2 liters

Procedure for preparation of panchagavya

Mix thoroughly fresh cow dung and cow ghee and incubate for two days. Add cow urine to the mixture and add 10 liters of water, stir properly in morning and evening. Incubate for 15 days. After 15 days of incubation add sugarcane juice, cow milk, cow curd, coconut water (crushed) and 15 ripe bananas. Incubate for 2 weeks.

Modified Panchagavya Formulation Experiment

H. Ramachandra Reddy and Bhaskara Padmodaya, UAS, Bangalore conducted an experiment to control soil borne Pathogen "*Fusarium oxysporum*" in tomato with MPG-3 and found that MPG-3 superior to carbendazim, MPG-3 decreased the plant disease index and increased vigour of plant and yield.

Natural Farming

Natural farming is based on the observation. It is about working with natural energies rather than the trying to conquer wild nature. It is distance from organic farming that is simply a return to the agriculture of the pre-chemical age. The problem of agriculture long pre-dates modern industrial farming methods. Everywhere farming has been widely practiced soils have been eroded and depleted and the natural biodiversity have been reduced. Understanding soil is central to natural farming. Soil is far from an inert substance, it is a complex living ecosystem comprising innumerable microorganisms that enable plant to take up nutrients essential for their growth and help defend them against disease and insects. These beneficial microorganisms include Rhizobium bacteria that convert atmospheric nitrogen into ammonium for use by plant, mycorrhizal fungi that help plant take up phosphorus and other nutrients, and many microbial pathogens that attack insect pests. Soil ecologists are just beginning to understand some of the complex interactions between soil microorganisms that enable nutrients to be retained and recycled, processes that are essential to the ecosystem above ground.

Masanobu Fukuoka is one of the pioneers of no-till grain cultivation was born on February 2, 1914. His system of farming is referred to as "natural farming", Fukuoka Farming or the Fukuoka Method. He wrote *The One-Straw Revolution, The Road Back to Nature* and *The Natural Way of Farming.*

Trained as a microbiologist in his native Japan, he began his career as a soil scientist specializing in plant pathology. At age 25, he began to doubt the wisdom of modern agricultural science. He eventually quit his job as a research scientist, and returned to his family's farm on the island of Shikoku in Southern Japan to grow organic mikans. From that point on he devoted his life to developing a unique organic farming system that does not require weeding, pesticide or fertilizer applications, or tilling.

The timing and circumstances of Fukuoka's conversion from Western agricultural science, parallels the new movement in the 1940s to organic farming and gardening in Europe and the US, led by pioneers like Lady Eve Balfour, Sir Albert Howard, and J.I. Rodale (founder of Rodale Press). However, it should be remembered that Fukuoka himself does not believe that he is an organic farmer:

"The problem, however, is that most people do not yet understand the distinction between organic gardening and natural farming. Both scientific agriculture and organic farming are basically scientific in their approach. The boundary between the two is not clear."

At 92, Fukuoka still manages to lecture when he can, as recently at the 2005 World Expo in Aïchi, Japan.

Technique

Fukuoka practices a system of farming he refers to as "natural farming." Although some of his practices are specific to Japan, the governing philosophy of his method has successfully been applied around the world. In India, natural farming is often referred to as "Rishi Kheti."

The essence of Fukuoka's method is to reproduce natural conditions as closely as possible. There is no plowing, as the seed germinates quite happily on the surface if the right conditions are provided. There is also considerable emphasis on maintaining diversity. A ground cover of white clover grows under the grain plants to provide nitrogen. Weeds are also considered as part of the ecosystem, periodically cut and allowed to lie on the surface so the nutrients they contain are returned to the soil. Ducks are let into the grain plot, and specific insectivorous carp into the rice paddy at certain times of the year to eat slugs and other pests.

The ground is always covered. As well as the clover and weeds, there is the straw from the previous crop, which is used as mulch, and each grain crop is sown before the previous one is harvested. This is done by broadcasting the seed among the standing crop. Also he re-introduced the ancient technique of seed balls (Tsuchi Dango {Earth Dumpling}). The seed for next season's crop is mixed with clay, compost, and manure then formed into small balls. Much less seed is used than in conventional growing, resulting in fewer but larger and stronger plants.

Herein lies the essence of Fukuoka's philosophy, and the source of it. Natural farming, as he defines it, is "Farming done as simply as possible, within and in cooperation with the natural environment, rather than the modern approach applying increasingly complex techniques to remake nature entirely for the benefit of human beings.

Natural farming is not as easy as it sounds, as Fukuoka himself learned the hard way. It is not easy, because in order to do natural farming one must first undo all the repercussions of years of so called "artificial" or scientific or hi-tech farming, which according to him is merely 'tampering with nature'. He found this at great cost of his father's fields.

Principles of natural farming

Natural farming is based on four principles, which are as follows.

- *No Cultivation:* It means no ploughing or turning of the soil. For centuries, farmers have assumed that the plough is essential for growing crops. However, noncultivation is fundamental to natural farming. The earth cultivates itself naturally by means of penetration of plant roots and the activity of micro-organisms, small animals and earhtworms.

- *No Chemical Fertilizer or Prepared Compost:* People interfere with nature, and try as they may, they cannot heal the resulting wounds. Their careless farming practices, drained the soil of essential nutrients, and the result is yearly depletion of the land. If left to itself, the soil maintains its fertility naturally, in accordance with the orderly cycle of plant and animal life.
- *No Weeding by Tillage or Herbicides :* Weeds play their part in building soil fertility and in balancing the biological community. As a fundamental principle, weeds should be controlled, not eliminated. Straw mulch, a ground cover of white clover interplanted with the crops, and temporary flooding provide effective weed control in my fields.
- *No Dependence on Chemicals :* From the time that weak plants developed as a result of such unnatural practices as ploughing and fertilizing, disease and insect imbalance became a great problem in agriculture. Nature left alone, is in perfect balance. Harmful insects and plant diseases are always present but do not occur in nature to an extent which requires the use of poisonous chemicals.

❑❑❑

8 GOOD AGRICULTURAL PRACTICES

Agricultural practices in India remained virtually unchanged for over 2000 years till 1950-60. With technological advancement in medicine and improvement in health care services, population in India substantially increased in the post independence period. The result was that there was acute shortage of food during 1960s and India had to import massive quantities of wheat from USA to avert famine. Most of these food shipments were in the form of aid. It was then that Government of India realised the gravity of situation and invested heavily in agricultural research. Fortunately, during this period food research laboratories in the West developed hybrid varieties of seeds to suit various types of soil and climatic conditions. At the same time new types of fertilizers and pesticides were introduced which showed dramatic increase in the yield. The agricultural research laboratories in India were able to build on this research and provided extension services to adopt new agricultural techniques to farms in India. This led to dramatic improvement in agricultural production which was hailed as "Green Revolution". Within two decades a country perennially deficient in food and subject to cyclic famines became a food surplus country in spite of the fact that there was a continuous rise in population which doubled in 30 years after independence. This trend continued into 1980's when Government built huge buffer stocks of wheat and rice to meet the needs during lean production years.

However the momentum of Green Revolution could not be maintained and certain disturbing trends were noticed in 1990s. The growth rate in productivity declined from 2.99 per cent per annum to 1.21 per cent This was not because we had come to a technical plateau our yields of various products are still quite low as compared to those achieved by developed countries and even some of the developing countries.

Main reason for low yields are out dated and unsustainable practices like excessive use of water together with imbalance in the use of fertilizers. The organic matter content in the soil has gone down because of less use of organic inputs and the micro-nutrient deficiencies have become alarming as brought out in the 10th plan document. This has adversely affected the soil health and environment such as depletion of water.

Keeping these developments in view, a National Agricultural Policy (NAP) 2000 was formulated which aims at 4% growth and relies on:

- Efficient use of resources
- Maximisation of benefits from exports of agricultural products, and
- Use of sustainable technology.

Broadly defined, a Good Agricultural Practice approach applies recommendations and available knowledge to addressing environmental, economic and social sustainability for on farm production and post-production processes resulting in safe and healthy food and non-food agricultural products. The concept of Good Agricultural Practice has evolved in recent years in this context to meet specific objectives of food security, food quality, production efficiency, livelihoods and environmental benefits at the local, national, regional and international levels.

- allowing for an acceptable level of pest damage;
- encouraging predatory beneficial insects to control pests;
- encouraging beneficial microorganisms
- careful crop selection, choosing disease-resistant varieties
- planting companion crops that discourage or divert pests;
- using row covers to protect crops during pest migration periods;
- rotating crops to different locations from year to year to interrupt pest reproduction cycles;
- Using insect traps to monitor and control insect populations.

The term best management practice (BMP) has been extensively used. Best management practice means a practice or combination of practices that are determined by a state after problem assessment, examination of alternative practices and appropriate public participation to be the most effective practicable means of preventing or reducing the amount of pollution generated by non point sources to a level compatible with water quality goals.

The categories of management practices for crop production includes;

a) *Crop management:* Tillage, crop sequence, seed improvement, etc.

b) *Soil and water management:* Runoff and erosion control, moisture conservation practices, wind erosion control, etc.

c) *Nutrient management:* Amount applied, methods of application, time of application, etc.

d) *Pest management:* Sanitation, pesticides usage, pest resistant crop, integrated measures, etc.

Each of these techniques also provides other benefits soil protection and improvement, fertilization, pollination, water conservation, season extension, etc.—and these benefits are both complementary and cumulative in overall effect on farm health. Effective organic pest control requires a thorough understanding of pest life cycles and interactions.

Criteria for defining best management practices

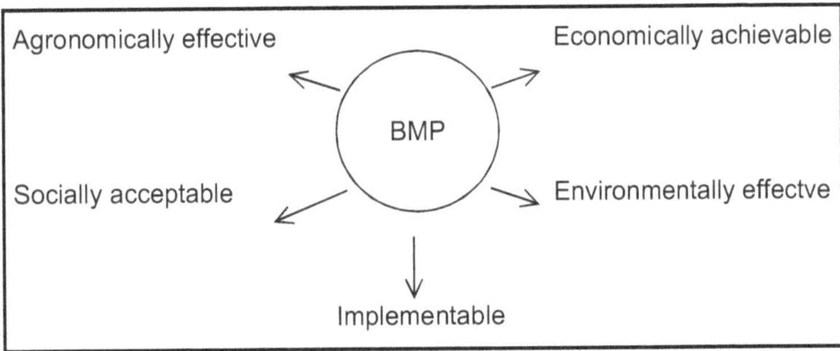

Agronomic effectiveness: Practices that promote productivity and commodity quality that conserve, preserve and if possible enhance soil productivity.

Environmental effectiveness: Practices that prevent, reduce or control pollutant loads.

Economically feasible: Practices whose economic implications are compatible with quality and competing goods.

Socially acceptable: An effective public participation process ensures that practices are reasonable and will have public support.

Implementability: This ensures that practices are legal, institutional existence is their or can be formulated in a way to provide flexibility and simplicity. With this background of GAP, certain agricultural practices are discussed below, which are based on the ecological principles of sustainable agriculture in achieving implementation.

Manure Management Systems

- Soil fertility and productivity would improve as major and micronutrients for crop growth and also a number of growth promoting substances are made available by compost-incorporation.
- The extremes of the pH of soil acidity, salinity, water logging etc., would be controlled by compost. Through these also, the nutrient availability would be improved.
- Compost and other organic materials curtail the leaching losses of nitrates.
- Use of compost with lignin and fibrous organic resources would result in slow degradation and helps in the formation of Humus. This helps in enhancing soil fertility and bringing stability in productivity.

Compost and biological activities

- With continuous use of compost, several useful micro-organisms and other biota like earthworms would multiply and their activities will degrade compost and release its energy. When compost is incorporated into soil, the carbon content of soil is raised and soil fertility is enhanced.
- Compost encourages nodulation in leguminous plant and promotes greater N2 fixation from the

atmosphere. It also helps release of fixed P from the soil. The efficacy of mycorrhiza in doing so is also increased.

- Soil borne diseases can be kept under check through increased microbial activities mediated by compost.
- Nurturing beneficial organisms and improving root growth of crops is effected by compost leading to greater absorption of nutrients and higher yields.

Thus, the role of compost through physical, chemical and biological activities in the soil is to enhance crop growth, augment productivity and improve the quality of crops.

Another method of augmenting the supplies of organic is the preparation of compost from farm house, and cattle-shed wastes of all types. Composting has been advocated and adopted extensively during the past 25 years. Composting is the process of reducing vegetable and animal refuse (rural or urban) to a quickly utilizable condition for improving the maintaining soil fertility. Research conducted in India and abroad has shown that good organic manure similar in appearance and fertilizing value to cattle manure can be produced from waste materials of various kinds, such as cereal straws, crop stubble, cotton stalks, groundnut husk, farm weed and grasses, leaves, leaf-mould, house refuse, wood ashes, litter, urine-soaked earth from cattle-sheds and other similar substances. These raw vegetables materials are rich in cellulose and other readily decomposable carbohydrates and have a carbon-nitrogen ratio of 40 or more to 1. The direct application of such undecomposed, low nitrogen organic matter as manure bring about a temporary deficiency of mineral nutrients (specially nitrates and ammonium compounds) in the soil by stimulating the growth of micro-organisms, which in turn, complete with crop plants for available nitrogen, phosphorus and other elements. Hence, before using them as manure, it is necessary to compost or partially decompose them. This process lowers the carbon-nitrogen ratio to about 10 or 12 to 1.

Composting

Composting is the breakdown of organic material by micro-organisms and soil fauna to give a humus end product called compost. It is an important technique for recycling organic waste (weeds, crop residues, waste from post harvest processing, dung, night soil, urine etc.) and for improving the quality and quantity of organic fertilizer. Compost is a slow release organic fertilizer which stimulates soil life and improves soil structure. It has also beneficial effects on the resistance of plants to pests and diseases.

Methods

Pit method

The site selected for the compost pit should be near to cattle shed and water source at high level so that no rain water gets during the monsoon. A temporary shed army be constructed over it to protect the compost from heavy rainfall. The pit should be about 1 m deep* 1.5 – 2.0 m wide and of any suitable length. The material brought from the cattle shed is spread and on each layer in spread a slurry of dung made with 4.5 kg urine each and 4.5 kg of inoculum taken from a 15 day old composting pit. A sufficient quantity of water (nearly 90%) is sprinkled over the material in the pit to the wet it. The pit is filled in this way layer-by-layer and it should not take longer than 1 week to fill. Care should be taken to avoid compacting the material in any way. The material is turned 3 times during the whole period of composting (i) after 15 days from filling the pit, (ii) another 15 days after, (iii) after another 30 days. At each turning the material is mixed thoroughly, moistened with water and replaced with the pit.

Heap method

During rainy seasons or in regions with heavy rainfall the compost may be prepared in heaps above ground. When sufficient nitrogenous material is not available a green manure or leguminous crop like sunhemp is grown on the fermenting heap before sowing seeds after the first turning. The green matters then turned in at the second mixing. The basic Indore pile is about 2 m wide at the base, 1.5 m high 2 m long. The sides are tapered so that the top is about 0.5 m narrower in width than the base. A small bund is sometimes built around the pile to protect it from wind which tends to deep the heap.

The heap is usually commenced with a 20 cm layer of carbonaceous material such as leaves, hay-straw, sawdust, wood chip and chopped corn stalks. This is then covered with the 10 cm of nitrogenous material such as fresh grass, weeds or garden plant residues, garbage, fresh or dry manure or digested sewage sludge. The pattern of 20 cm carbonaceous material and 10 cm nitrogenous material is followed until the pile is 1.5 m high and they are normally wetted so that they feel damp but not soggy. The pile is sometimes covered with soil or hay to retain heat and is turned at 6 and 12 weeks interval.

Bangalore method

Trenches or pits 1 m deep are dug; the breadth and length of the trenches can be made depending on the availability of land and the type of material to be composted. The selection of site for 1 pit is made as mentioned in the Indore method. The trenches should preferably have stopping walls and a floor of 90 cm slope to prevent waterlogging. Organic residues and night soil are put in alternate layers and after filling, the pit is covered with a 15-20 cm thick layer of refuse, the materials are allowed to remain in the pit without turning and watering for 90 days. The methods requires along time to produce a finished compost.

Synthetic method

In the preparation of synthetic compost, the organic nitrogen as dung, required by micro-organism, can be completely substituted by inorganic nitrogen compounds like ammonium sulphate and urea which are utilized equally effectively for decomposition of carbonaceous materials into compost. This facilitates the utilization of large quantities of various organic waste material where supplies of dung are either short of the requirement or not available at all, as on mechanized farms.

The material to be composted is moistened and is sprinkled with the fertilizer solution and then with lime. Superphosphate may be added to fortify the phosphorous content of the manure. The treatment is continued layer-wise until the heap or pits is filled to the size and allowed to ferment. The manure becomes ready for application in about 120-180 days and resembles with farmyard manure in its action on soil and plant growth.

Japanese method of composting

Preparation of vats from stone-slabs or bamboo/ lantana (Shiva Shankar, 1996).

In Japanese Method, instead of pits, vats from granite stone slabs are made. They measure 18′ to 30′ in length, 3′ to 4′ in width and have a height of 2.5′ to 3.0′. The walls of the stone slabs are so constructed to allow for easy aeration as in MOA models seen in photographs. They may also have holes or windows. The bottom of the vats are covered by stone-slabs and the joints and cracks are plastered with cement to prevent leaching of nutrients to soil below. The cost for such a vat would be Rs. 5,000/- to Rs.8,000/- and the investment is only for one time. If this is not feasible, vats can be fabricated with the same dimensions from locally available materials. Stakes of about 2.5′ to 3′ can be fixed on all sides of the proposed vat and a fence constructed around this with bamboo mats or lantana branches. For the ground, a hard non-leaky surface can be prepared with broken stones and brick bats and plastered with cement, lime, clay and cowdung etc.

Advantages from the vats

In the Japanese method, vats play an important role. Unlike development of high temperature (65^0 to 70^0 C) and loss of nitrogen through leaching of nitrates in the ordinary pit system, these are prevented in the new method. Besides, being on the surface of the ground, it is easy to turn the organic residues regularly.

Materials suited for compost preparation

Principally, compost can be prepared from five organic waste materials.

a) Cowdung, poultry wastes etc.

b) Grass, dry and green leaves, crop residues and weeds also.

c) Wastes from agro based industries (Pressmud, vegetable and fruit wastes from processing units, sericultural wastes etc.)

d) Naturally occurring nutrient source (Rock Phosphate)

e) Organic wastes from towns and cities and from human being and treated wastes of sewage and sludge.

Micro-organisms to convert organic wastes

In order to convert and degrade varied sources of organic wastes in vats, a number of microbial cultures have to be adopted. Micro-organisms viz., *Aspergillus, Penicellium* and *Trichoderma viride* can degrade grass, straw and stover and improve the status of nitrogen and phosphorus and form humus. Agrobacterium, Radiobacter, Azotobacter and Bacillus can all be used in composite cultures which can enrich nitrogen. Coir pith and others need Pleurotus and Polyporus.

Preparation for composting

All materials required should be collected and assembled in order. Organic wastes of coconut shells with coir, coconut branches and leaves, dry leaves, grass, straw, stover and residues should be cut into small pieces and kept separately. Cowdung, poultry waste, biogas slurry etc., be kept ready. Rock Phosphate, gypsum and lime be weighed and in he right form. Packets of microbial cultures, samples of old compost, threshed waste or husk of paddy or ragi, jaggery, water and plastic bucket are all required for preparing for the spread of microbes.

Compost making

Varied organic wastes are to be spread layer by layer. Yet one small portion of about 2 space be kept free on one side of the vat to facilitate turning in the residues regularly. Sliced and broken pieces of coconut shells, leaves, fibrous material, tender tree barks or pieces, wood ash, etc., should form the bottom layer of 10-15 cm. Height. These do not degrade easily as they contain lignin. Second layer of dry leaves, grass, residues, groundnut haulms etc., can be of 10-15 cm. These two hard and dry layers absorb moisture and nutrients leached from the top. Cowdung, urine and biogas slurry can be used singly or mixed in a bucket of water and sprinkled over. A very small sprinkling of soil and ash can be spread over this. In the third layer, green leaves of Pongamia, Albizzia (Bage), green grass and weeds. Dhaincha, *Sesbania* and crop residues can form 10-15 cm. These are rich in nitrogen. The fourth layer should contain organic wastes rich in phosphorus. Roots of green manures and legumes, Acalypha, rock phosphate etc., can be used. In the same layer, potash-rich items like Calatropis, Datura and other weeds, tomato and tobacco crop residues, ash, poultry wastes etc., can form 10-15 cm. Over these, two to three buckets of cowdung and biogass slurry can be sprinkled. This should seed down also to wet the lower surfaces. Generally, while preparing farm yard manure, paddy or ragi straw is also seen. If we have to derive greater benefits, these straw and stover samples should be cut into very thin pieces of 4 to 5 cm and should form a separate layer. In this case, the fifth layer. These are rich in carbon and would provide energy to micro organisms they also have the nutrient 'Silicon'. If green biomass is available along with dry straw, they can also be mixed in this small layer.

The sixth layer should exclusively contain 20-30 cm of cowdung. If stored in a heap for a long time, part of it would have dried up. These should be powdered. Over theses, in small quantities, old compost powder, tank-silt or soil and ash can be sprinkled.

Treatment with microbial cultures

Depending on varied organic biomass taken for composting, suitable microbial cultures or composite cultures have to be mixed in water and paddy husk or threshed material soaked in this liquid and the whole spread uniformly. Microbial cultures can be obtained from the Departments of Microbiology of Agricultural Universities or from Biofertilizers companies. Thus, sprinkling of cultures could be one each layer or at the end of sixth layer as presented here. If this is not possible, old compost powder and 4 to 5 percent jaggery in water could be used. A thin layer of soil or cowdung / slurry can be spread over the culture-treatment. These six major layers form one 'set' or a group of layers. The second and third sets need not have coconut shreds and leaves etc.

After all sets are filled, care should be taken to see that moistures should not exceed 65 to 70 percent and temperature should not raise very high. It is necessary to turn the sets after 15 days after filling for the first time and every month thereafter. Sometimes a small amount of lime can help in quick decomposition of hard materials. To enrich compost, 50 to 100 kg. rock phosphate per 100 kg. compost and a similar quantity of gypsum can be added along with some oil cakes which are not used for feeding cattle during the last turning of the compost.

The quality of compost is dependent on the availability or organic wastes and their composition besides their judicious mixing and preparation

Crop Rotation

Crop rotation is a planned order of specific crops planted on the same field. Crop rotation also means that succeeding crops are of a different genus, species, subspecies, or variety than the previous crop. Crop rotation is an easy way to control diseases and insects at no cost. For example, tomatoes, cauliflower or cabbage planted in the same location each year will actually encourage buildup of certain diseases in the soil. By rotating crops, you are removing the host plant and preventing the spread of disease.

Encyclopaedia Britannica Article defines crop rotation as the successive cultivation of different crops in a specified order on the same fields, in contrast to a one-crop system or to haphazard crop successions.

Crop rotation is the sequence of cropping where two dissimilar type of crops follow each other - a few examples include cereals and legumes, deep-rooted and short-rooted plants and where the second crop can make use of the manuring or irrigation provided some months earlier to the first crop. The combinations possible are endless, and will depend to a great deal on the local situations.

Crop rotation is the practice of growing a series of dissimilar type of crops in the same space in sequential seasons to avoid the buildup of pathogens and pests that often occurs when one species is continuously cropped. Crop rotation also seeks to balance the fertility demands of various crops to avoid excessive depletion of soil nutrients. A traditional component of crop rotation is the replenishment of nitrogen through the use of legumes in sequence with cereals and other crops. It is one component of polyculture. Crop rotation can also improve soil structure and fertility by alternating deep-rooted and shallow-rooted plants.

Urdbean is grown in summer and Kharif in northern India and Kharif and rabi in sourthern Indian conditions. It may, therefore, be included in several crop rotations. The important rotations are given as under:

- Blackgram-toria-wheat
- Pigeonpea+urdbean-wheat
- Sorghum/maize+urdbean-wheat
- Maize-mustard-urdbean
- Maize-potato-urdbean
- Rice-urdbean
- Urdbean-toria-wheat-mungbean
- Maize-urdbean-wheat-green manuring
- Maize-safflower
- Rice-chickpea
- Blackgram-mustard
- Groundnut-safflower
- Rice-mustard
- Rice-fieldbeans
- Rice-groundnut
- Soybean-wheat
- Fingermillet-chickpea
- Sunflower-chickpea
- Jute-wheat

Examples would be barley after wheat, row crops after small grains, grain crops after legumes, etc. The planned rotation sequence may be for a two- or three-year or longer period. Some of the general purposes of rotations are to improve or maintain soil fertility, reduce erosion, reduce the build-up of pests, spread the workload, reduce risk of weather damage, reduce reliance on agricultural chemicals, and increase net profits.

Rotating to a different crop such as wheat on barley ground usually results in higher grain yields when compared to continuous cropping of wheat. Even greater benefits are usually obtained by rotating two distinctly unrelated crops, such as a small grain seeded into land where the previous crop was a legume or other herbaceous dicot such as flax or sunflower. Many of the reasons for the beneficial effects of rotations are not completely understood.

Some of the more important beneficial effects that can be obtained from a well planned crop rotation are:

- Reduced insect and disease problems
- Beneficial residual herbicide carryover
- Improved soil fertility
- Improvements in soil tilth and aggregate stability
- Soil water management
- Reduction of soil erosion
- Reduction of allelopathic or phytotoxic effects

Pest control is often an important reason for crop rotation. Rotations can be used to prevent or partially control several pests and reduce the reliance on chemical and mechanical control. A combination of crop rotation and pesticides is often more effective in reducing pest populations to economic levels than pesticides alone. Pesticides that provide economical control are not available for pests such as white mold in potato, dry bean, and sunflower, and crop rotations are the only feasible control method for reducing the impact of sclerotinia.

Each crop has different fertilizer requirements. By changing the location of your crops you can avoid the risk of depleting the soil of specific nutrients. Some crops will actually add essential elements to the soil. By using crop rotation, you can actually build up the soil over the years.

History

Crop rotation was already mentioned in the Roman literature, and referred to by great civilizations in Africa and Asia. From the end of the Middle Ages until the 20th century, the three-year rotation was practised by farmers in Europe with a rotation: rye or winter wheat, followed by spring oats or barley, then letting the soil rest (fallow) during the third stage. The fact that suitable rotations made it possible to restore or to maintain a productive soil has long been recognized by spring crops, in place of grains for human consumption.

Throughout human history, wherever food crops have been produced, some kind of rotation cropping appears to have been practiced. One system in central Africa employs a 36-year rotation; a single crop of finger millet...

A four-field rotation was pioneered by the Dutch and popularised by the British agriculturist Charles Townshend in the 18th century. The system (wheat, barley, turnips and clover), opened up a fodder crop and grazing crop allowing livestock to be bred year-round. The four field rotation was a key development in the British Agricultural Revolution.

Crop rotation was pioneered in the USA by George Washington Carver, not only through laboratory work, but through practical promotion and education.

In the Green revolution, the practice of crop rotation gave way in some parts of the world to the practice of simply adding the necessary chemical inputs to the depleted soil, e.g. replacing organic nitrogen with ammonium nitrate or urea and restoring soil pH with lime. However, disadvantages of monoculture from the standpoint of sustainable agriculture have since become apparent.

Crops are changed year by year in a planned sequence. Crop rotation is a common practice on sloping soils because of its potential for soil saving. Rotation also reduces fertilizer needs, because alfalfa and other legumes replace some of the nitrogen corn and other grain crops remove.

Method and purpose

Crop rotation avoids a decrease in soil fertility, as growing the same crop repeatedly in the same place eventually depletes the soil of various nutrients. A crop that leaches the soil of one kind of nutrient is followed during the next growing season by a dissimilar crop that returns that nutrient to the soil or draws a different ratio of nutrients, for example, rice followed by cotton. By crop rotation farmers can keep their fields under continuous production, without the need to let them lie fallow, and reducing the need for artificial fertilizers, both of which can be expensive.

Legumes, plants of the family Fabaceae, for instance, have nodules on their roots which contain nitrogen-fixing bacteria. It therefore makes good sense agriculturally to alternate them with cereals (family Poaceae) and other plants that require nitrates. A common modern crop rotation is alternating soybeans and maize (corn). In subsistence farming, it also makes good nutritional sense to grow beans and grain at the same time in different fields.

Crop rotation is also used to control pests and diseases that can become established in the soil over time. Plants within the same taxonomic family tend to have similar pests and pathogens. By regularly changing the planting location, the pest cycles can be broken or limited. For example, Root-knot nematode is a serious problem for some plants in warm climates and sandy soils, where it slowly builds up to high levels in the soil, and can severely damage plant productivity by cutting off circulation from the plant roots. Growing a crop that is not a host for Root-knot nematode for one season greatly reduces the level of the nematode in the soil, thus making it possible to grow a susceptible crop the following season without needing soil fumigation.

This principle is of particular use in organic farming, where pest control and may be achieved without synthetic pesticides.

The choice and sequence of rotation crops depends on the nature of the soil, the climate, and precipitation which together determine the type of plants that may be cultivated. Other important aspects of farming such as crop marketing and economic variables must also be considered when choosing a crop rotation.

The aim of rotation is threefold: to balance nutrient demands, foil insect and disease attacks, and deter weeds.

Nutrient Rotation

i) The challenge here is trying to balance the nutrient demands each crop makes on the soil.

ii) Divide crops into the following four types, for four different seasonal rotations:

- *Leaves:* Thrive on nitrogen; examples include lettuce, salad greens, chicory, spinach, broccoli, cabbage, cauliflower,
- *Fruits:* Need phosphorus; examples include cucumbers, melons, pumpkins, tomatoes, chilli, and brinjal.
- *Roots:* Love potassium; examples include onions, garlic, carrots, beets, turnips, and radishes.

iii) *Soil builders and cleaners:* Legumes are excellent for the soil because they store nitrogen from the air and release it into the soil; examples of cleaners include corn and potatoes, examples of builders include beans and peas.

The first season of planting could be devoted to leafy plants, the next season to fruits, followed by the root plants and then legumes.

Prevention rotation

Crop rotation can also break the cycles of pest and disease problems that build up in soils planted repeatedly to the same crop.The idea is to plan your rotation so that no two crops subject to similar diseases follow one another within the disease's incubation period. The same principle holds for insect pests: crop rotation makes it harder for emerging insects to find their preferred food each spring.

- Don't follow tomato, chilli or brinjal with potatoes, or each other.
- Allow 3 years before replanting the same group in any given bed.
- Onions may be planted throughout all groups.
- Beets, carrots and radishes may be planted among any group, and replanted as early crops are removed.
- Don't forget to interplant with companion plants to minimize pesticide use. See the Companion Plants handout for some ideas on this practice.

Weed suppression

To help build organic matter, you might also consider using a "green manure" sometimes called a cover crop. There are both summer and winter cover crops which will act as smoother crops.

By adding organic matter in this way, you will increase aeration and water holding capacity of your soil, prevent weed growth and soil erosion, and support the beneficial organisms necessary for a healthy, living soil.

Polyculture

Polyculture is one of the principles of permaculture. It is agriculture using multiple crops in the same space, in imitation of the diversity of natural ecosystems, and avoiding large stands of single crops, or monoculture. It includes crop rotation, multicropping, intercropping, companion planting, beneficial weeds and alley cropping.

Several polycultures, such as cassava-beans-maize, cassava-tomato-maize, and sweet potato-maize, were tested. Productivity evaluation of these polycultures indicates 2.82, 2.17 and 1.45 times greater productivity than monocultures respectively. The use of *Crotalaria juncea* and *Vigna unguiculata* as green manure has ensured a production of squash equivalent to that obtainable by applying 175 kg/ha of urea. In addition, such legumes improved the physical and chemical characteristics of the soil and effectively broke the life cycles of insect pests such as the sweet potato weevil.

Advantages

Polyculture, though it often requires more labour, has several advantages over monoculture:

- The diversity of crops avoids the susceptibility of monocultures to disease. For example, a study in China reported in Nature showed that planting several varieties of rice in the same field increased yields by 89%, largely because of a dramatic (94%) decrease in the incidence of disease, which made pesticides redundant.

- The greater variety of crops provides habitat for more species, increasing local biodiversity. This is one example of Reconciliation Ecology, or accommodating biodiversity within human landscapes. The science of agroecology has revealed the benefits of polyculture (multiple crops in the same space), which is often employed in organic farming. Planting a variety of vegetable crops supports a wider range of beneficial insects, soil microorganisms, and other factors that add up to overall farm health, but managing the balance requires expertise and close attention. Polyculture is using multiple crops which are beneficial to each other. The output of one crop becomes the input of another thus creating a balance in the soil and environment. Polycultures, as opposed to monocultures, grow two or more crops together, not just one.

Kovach's four designs, even more diverse than typical polycultures, combine apples, peaches, green beans, tomatoes, strawberries, blueberries, raspberries and soybeans. But each design tests a different arrangement. The first has solid rows, with each row having a single crop, and the crop height switching from row to row: for example, a row of high trees, a row of low strawberries, a row of high trees, a row of low tomatoes.

The second design mixes more than one crop within a row but keeps the high crops and low crops together in their own rows. Apples, peaches and raspberries, for example, would line up in a row, then green beans, strawberries and tomatoes in the next, as a way to roadblock infestations.

"The concept, is that insect pests seem to move down rows. So if you're an apple pest, you might stop at the peaches. A peach pest might stop at the raspberries. A raspberry pest at the blueberries. And so forth."

The third design goes a step further. It mixes the crops within a row and also alternates heights in the row. A single row might grow apples then strawberries, peaches then green beans, raspberries then tomatoes. "checkerboard" system.

The fourth design adds raised beds to the equation with mixed rows planted within.

All four designs employ drip irrigation, disease-tolerant and -resistant varieties, fencing against rabbits and woodchucks, staggered planting dates for the annuals and maturity dates for the perennials (allowing for early, mid- and late-season harvest and season-long production), and newer, less-toxic pesticides if and as needed, with sustainability, not 100 per cent organic production, the goal.

Food production has always been primarily polycultural, from primitive hunting and gathering to current third world house gardens and managed agroforests. For 98.5% of farming history, humans have produced food from integrated polycultures. In both the first and third world, polyculture practices remain important and the majority of current world farmers depend on multi-species production for their livelihood. Evidence emerged of the growing interest in polyculture indicated by the increasing amount and diversity of the literature.

Multiple Cropping

In agriculture, multiple cropping is the practice of growing two or more crops simultaneously in the same space during a single growing season. It is a form of polyculture. A related practice, companion planting, is sometimes used in gardening and intensive cultivation of vegetables and fruits. One example of multi-cropping is tomatoes + onions + marigold; the marigolds repel some of tomato's pests.

Multiple cropping is the practice of planting several different crops on the same plot of land at the same time. It is common among small-scale farmers in Africa. Crops, livestock and trees can all be integrated into a small farm, making it much more productive. Such integrated farming is particularly relevant for small farms or fields.

The integration of many farm enterprises gives farm families several advantages. More crops can be planted in a small space. For example, intercropping and relay cropping can allow a farmer to plant two crops maize and beans, for example in the field at the same time. The production of crops is usually spread over a longer period of the year, allowing for better vegetative cover to protect the soil, but also spreading out the harvest throughout the year.

The appropriate crops, crop combinations, planting times and planting patterns will vary from place to place, depending on the local climate, soils, topography, water availability, pests and diseases, socio-economic conditions, and other factors.

Advantages

- Multiple cropping reduces the risk of total loss from drought, pests and diseases. Usually at least some of the crops can escape disaster and produce a yield.
- It optimizes production from small plots, so can help farmers cope with land shortages.
- Including legumes in the cropping pattern helps maintain soil fertility by fixing nitrogen in the soil.
- Multiple cropping yields different types of produce, resulting in a balanced diet for the family.
- It suppresses weeds. As the planting density is high, weeds cannot compete with the crops.

- Different types of crops can be planted to take advantage of different seasons. For example, crops that require a lot of water can be grown in the wet season, intercropped with drought-resistant crops that can be harvested in the following dry season.

Disadvantages

- The presence of crops in the field throughout the year allows crop pests to survive more easily. Some pests can shift from one crop to another: for example, aphids can move to cotton plants during the dry season.
- The large number of different crops in the field makes it difficult to weed.

It may be difficult to introduce new technologies such as row planting, modern weeding tools, and improved varieties.

Multiple cropping is found in many agricultural traditions. In the Garhwal Himalaya of India, a practice called *baranaja* involves sowing 12 or more crops on the same plot, including various types of beans, grams, and millets, and harvesting them at different times.

In the cultivation of rice, multiple cropping requires effective irrigation, especially in areas with a dry season. Rain that falls during the wet season permits the cultivation of rice during that period, but during the other half of the year, water cannot be channeled into the rice fields without an irrigation system. The Green Revolution in Asia led to the development of high-yield varieties of rice, which required a substantially shorter growing season of 100 days, as opposed to traditional varieties, which needed 150 to 180 days. Due to this, multiple cropping became more prevalent in Asian countries.

One kind of multiple cropping is intercropping, where an additional crop is planted in the spaces available between the main crop. The other kind is sequence cropping in which the succeeding crop is sown after the harvest of preceding crop.

Multiple cropping zones for rain-fed crop production

To assess the multiple cropping potential, a number of multiple cropping zones have been defined through matching both growth cycle and temperature requirements of individual suitable crops with time available for crop growth. For rain-fed conditions this period is approximated by the LGP, *i.e.,* the number of days during which both temperature and moisture conditions permit crop growth.

For the definition of multiple cropping zones four types of crops are distinguished: thermophilic crops requiring warm temperatures, cryophilic crops performing best under cool and moderately cool conditions, hibernating crops, and wetland crops with specific water requirements. Furthermore, the crops are subdivided according to growth cycle length, namely of less or more than 120 days duration, respectively. According to the above criteria, the following nine zones were classified and mapped:

a) *Zone of no cropping* (too cold or too dry for rain-fed crops)

b) *Zone of single cropping*

c) *Zone of limited double cropping* (relay cropping; single wetland rice may be possible)

d) *Zone of double cropping* (sequential cropping; wetland rice not possible)

e) *Zone of double cropping* (sequential cropping; one wetland rice crop possible)

f) *Zone of limited triple cropping* (partly relay cropping; no third crop possible in case of two wetland rice crops)

g) *Zone of triple cropping* (sequential cropping of three short-cycle crops; two wetland rice crops possible)

h) *Zone of triple rice cropping* (sequential cropping of three wetland rice crops possible)

Words relating to multiple cropping

Agroforestry

Growing trees along with annual crops and livestock. Here perennial woods components are deliberately grown along with herbaceous component.

Alley cropping

Growing annual crops between rows of (often leguminous) trees or shrubs. Prunings from the trees or shrubs can be used as fertilizer, mulch or livestock fodder. Alley refers to the space in between hedge rows.

Cropping pattern

The yearly sequence and spatial arrangement of crops, or of crops and fallow, on a given area.

Cropping system

The cropping patterns used on a farm and their interaction with farm resources, other farm enterprises and available technology which determine their make-up.

Farming system

All the elements of a farm which interact as a system, including people, crops, livestock, other vegetation, wildlife, the environment and the social, economic and ecological actions between them.

Intercropping

Growing two or more crops in the same field at the same time. Crops can be planted in rows (row intercropping), or the seed can be mixed and broadcast (mixed intercropping). Planting in rows makes applying fertilizer, weeding and harvesting easier.

Monoculture

Growing the same crop year after year on the same piece of land.

Multiple cropping

Growing two or more crops in the same field in one year at the same time, or one after the other. Multiple cropping can be done with annual food crops, biennial crops (such as cotton), fodder crops, and tree crops.

Relay cropping

Growing two or more crops in a field with their growing seasons overlapping: e.g., planting a second crop in the field where another is already growing. After the harvest of the first crop, the second (often drought-resistant) crop continues to grow, and is harvested later.

Rotation

Changing the crops grown on a particular piece of land from season to season (or changing from crops to fallow).

Sequential cropping

Growing two or more crops in sequence in the same field in the same year. The second crop is planted after the first one is harvested.

Sole cropping (solid planting)

One crop variety grown alone in a pure stand.

Strip cropping

Planting of alternate strips of grasses or grains with other crops along the contour in order to conserve moisture and decrease erosion.

Vermicomposting

Vermicompost

Vermicompost (also called Worm Compost, Vermicast, or Worm Manure) is end product of the breakdown of organic matter by special varieties of earthworms. Organic and the process is known as vermicasting. The product is the result of organic waste consumed by earthworm, digested and excreted in the form of granules.Vermicompost is a nutrient-rich, natural fertilizer and soil conditioner. The process of producing vermicompost is called *vermicomposting*.

Vermiculture is a branch deals with rearing and maintaining of earthworms for vermicompost preparation. Vermiculture means artificial rearing or cultivation of worms (Earthworms) and the technology is the scientific process of using them for the betterment of human beings. Vermicompost is the excreta of earthworm, which is rich in humus. Earthworms eat cow dung or farm yard manure along with other farm wastes and pass it through their body and in the process convert it into vermicompost. The municipal wastes; non-toxic solid and liquid waste of the industries and household garbage's can also be converted into vermicompost in the same manner. Earthworms not only convert garbage into valuable manure but keep the environment healthy. Conversion of garbage by earthworms into compost and the multiplication of earthworms are simple process and can be easily handled by the farmers.

In addition to worms, a healthy vermicomposting system hosts many other organisms such as insects, molds, and bacteria. Though these all play a role in the composting process, the earthworm is the major catalyst for the composting process.

Generally there are 3200 species of earthworms are occurring in nature. Out of these *Eisenia fetida* and *Eudrelis eugina* are used for preparing vermicomposed.

Methods

Multiplication of worms in large scale

Prepare a mixture of cow dung and dried leaves in 1:1 proportion. Release earthworm @ 50 numbers/10 kg. Of mixture and mix dried grass/leaves or husk and keep it in shade. Sprinkle water over it time to time to maintain moisture level. By this process, earthworms multiply 300 times within one to two months. These earthworms can be used to prepare vermicompost.

Two methods are used for vermicompost preparation

- Heap method
- Trench method

Method of preparation of vermicompost

Vermicomposting methodology

A thatched roof shed preferably open from all sides with unpaved (katcha) floor is erected in East-West direction length wise to protect the site from direct sunlight. A shed area of 12′x12′ is sufficient to accommodate three vermibeds of 10′x3′ each having 1′ space in between for treatment of 9-12 quintals of waste in a cycle of 40-45 days. The length of shed can be increased/decreased depending upon the quantity of waste to be treated and availability of space. The height of thatched roof is kept at 8 feet from the centre and 6 feet from the sides. The base of the site is raised at least 6 inches above ground to protect it from flooding during the rains. The vermibeds are laid over the raised ground as per the procedure given below.

The site marked for vermibeds on the raised ground is watered and a 4″-6″ layer of any slowly biodegradable agricultural residue such as dried leaves/straw/sugarcane trash etc. is laid over it after soaking with water. This is followed by 1″ layer of Vermicompost or farm yard manure.

Earthworms are released on each vermibed at the following rates :

For treatment of cowdung/agriwaste : 1.0 kg.

For treatment of household garbage : 1.5 kg.

The frequency and limits of loading the waste can vary as below depending upon the convenience of the user

Frequency	Loading
Daily	2" /bed/day
In Bulk	12-15"(3-4q/bed/cycle of 45 days)

The loaded waste is finally covered with a Jute Mat to protect earthworms from birds and insects. Water is sprinkled on the vermibeds daily according to requirement and season to keep them moist. The waste is turned upside down fortnightly without disturbing the basal layer (vermibed). The appearance of black granular crumbly powder on top of vermibeds indicate harvest stage of the compost. Watering is stopped for at least 5 days at this stage. The earthworms go down and the compost is collected from the top without disturbing the lower layers (vermibed). The first lot of Vermicompost is ready for harvesting after 2-2½ months and the subsequent lots can be harvested after every 6 weeks of loading. The vermibed is loaded for the next treatment cycle.

Steps to be followed for vermicompost preparation

- The basin of the tree itself can be used as a vermibed.
- Vermicastings at the rate of about 5-10 kgs. should be applied per tree depending on the size and age of the tree.
- About 25 kgs. of any farm yard manure (FYM) should be applied evenly on the top of the vermnicatings.

- This is then mulched with organic litter. Slashed weeds available on the farm can be used for this purpose.
- Watering can be done as per the regular applications itself, ensuring that proper moisture is maintained.
- When all the above steps are followed, a conducive atmosphere is created for triggering the vermicastings and starting the vermiculture process.
- Once the earthworms have the suitable environment for existence, they start consuming the organic matter and turning it into rich vermicompost.
- This vermicompost is a bio-fertilizer enriched with beneficial soil micro-organisms.
- The vermicastings are highly stable and do not disintegrate thus preventing soil erosion.
- The vermicompost contains all the essential plant nutrients like N, P and K, thus eliminate usage of any further chemical inputs.

Vermicomposting using paddy straw

Vermicomposting is an appropriate technique for efficient recycling of animal wastes, crop residues and agro-industrial wastes. Paddy straw is a wide C: N (80:1) organic material, low in nitrogen and phosphorus but fairly rich in potassium. In conventional method of composting, paddy straw takes 6-8 months for decomposition resulting in a poor quality of compost. The process of conversion of organic materials into manure is chiefly microbiological and greatly influenced by the proportion of carbonaceous and nitrogenous materials present in organic wastes.

Microorganisms need carbon for cell structure formation and nitrogen for cellular protein synthesis. It was found that C: N ratio of 30:1 or lower for raw material was desirable for efficient composting. So, C: N ratio of organic materials poor in nitrogen should be made narrow by adding nitrogen in the form of any nitrogenous fertilizer to it for better decomposition. Super-phosphate is generally added to fortify the phosphorous content of the compost.

When beginning a vermicomposting bin, add as many composting worms as available. They should be added to moist bedding. Quantities of kitchen waste appropriate for the worm population can be added to the bin daily or weekly. At first, feed the worms approximately ½ their body weight in kitchen scraps a day, maximum. After they have established themselves, you can feed them up to their entire body weight.

Bedding

Bedding in a worm bin is the living medium for the worms but is also used as a food source, it is material that is high in carbon and is made to mimic dried leaves on the forest floor, which is the worms' natural habitat. The bedding needs to be moist (often related to the consistency of a wrung-out sponge) and loose to enable the earthworms to breath and to facilitate aerobic decomposition. A wide variety of bedding materials can be used including newspaper, sawdust, hay, cardboard, peat moss, aged manure (meaning the manure has to be pre-composted before use), and dried leaves. Most vermicomposters avoid using any glossy papers from newspapers and magazines, junk mail and shredded paper from offices, because they may contain toxins which will severely affect the system. Also some cardboard cannot be used if it contains wax or plastic, such as cereal boxes, and other boxes designed to hold food items. Newspapers and phone books printed on regular, non-glossy pages are heavily regulated by the FDA and use non-toxic soy and Canola based inks.

Temperature

The worms that are used in composting systems prefer temperatures between 55 and 70 degrees Fahrenheit (12-21°C), and temperature of the bedding should not get below freezing or above 85 degrees Fahrenheit (29° C).

Kitchen waste

Greens : If too much kitchen waste is added for the worms to process, the waste will putrify. A balance between "green matter" such as kitchen scraps and "brown matter" such as shredded newspaper for bedding must be maintained in order for the worms to do their work. This balance should be approximately one part "green matter" for every two parts of "brown matter". Covering the kitchen scraps with a layer of "brown matter" also has the added benefit of reducing odor and insect problems. Avoid grass clippings or other plant products that have been sprayed with pesticides. In a small bin, this includes banana peels which can kill everything in the bin, if heavily sprayed.

Over the long term, care should be taken to maintain optimum moisture levels and pH balance. In a non-continuous-flow vermicomposting bin, excess liquid can be drained via a tap and used as plant food. A continuous flow bin will not retain excess liquid and it requires extra water to be added to keep the bedding moist. It is commonly believed that too many citrus peels in the material to be composted can cause an intelorable level of acidity, which can be mitigated by adding an occasional handful of lime. It is more likely to be the chemical d-limonene (best known for being the juice that spurts out when an orange is peeled) which affects worms.

Worms as well as other microorganisms in the composting process require oxygen, so the bin must "breathe". This can be accomplished by regularly removing the composted material, adding holes to a composting bin, or using a continuous-flow bin. If insufficient oxygen is available, the compost will become anaerobic. This will provide a host environment for a different type of decay process which produces a strong odor offensive to most people. This type of decay is found in swamps and bogs and is responsible for the stench sometimes found in these environments.

Feeding methods

There are basically two methods of adding more matter to the bin.

- The first method, known as top feeding, is when organic matter placed directly on top of the existing layer bedding in a bin and then covered with another layer of bedding. This is repeated every time the bin is fed.
- The other method of feeding is known as pocket feeding. In this method a top layer of bedding is maintained and food is buried beneath. The location of the food is changed each time and often the bin is fed in more than one location. As bedding runs low more is added. Vermicomposters often use a combination of both methods. Sometimes by not burying the food it can attract fruit flies.

Advantages

Vermicompost properties

Worm compost is usually too rich for use as a seed compost, but is useful as a top layer of soil or an addition to potting composts. Some types of pitted seeds are reportedly easier to germinate when placed in vermicompost for several months.

Vermicompost is beneficial for soil in three ways

- It improves the physical structure of the soil.
- It improves the biological properties of the soil (enrichment of micro-organisms, addition of growth hormones such as auxins and gibberellic acid, and addition of enzymes, such as phosphates, cellulase, etc.).
- It attracts deep-burrowing earthworms already present in the soil.

Contents	
Nitrogen	3-5%
Phosphorous	0.3%
Potassium	0.56%
Micro nutrients	Traces
Organic Matter	40-60%

Advantages of Vermicomposting

- Vermicompost is an ecofriendly natural fertilizer prepared from biodegradable organic wastes and is free from chemical inputs.
- It does not have any adverse effect on soil, plant and environment.
- It improves soil aeration, texture and tilth thereby reducing soil compaction.
- It improves water retention capacity of soil because of its high organic matter content.
- It promotes better root growth and nutrient absorption.
- It improves nutrient status of soil-both macro-nutrients and micro-nutrients.

Nutrient Profile of Vermicompost and Farm Yard Manure

Nutrient	Vermicompost	Farm Yard Manure
N (%)	1.6	0.5
P_2O_5 (%)	0.7	0.2
K_2O (%)	0.8	0.5
Ca (%)	0.5	0.9
Mg (%)	0.2	0.2
Fe (ppm)	175.0	146.5
Mn (ppm)	96.5	69.0
Zn (ppm)	24.5	14.5
Cu (ppm)	5.0	2.8
C:N ratio	15.5	31.3

Source: Punjab State Council for Science and Technology, Chandigarh

Benefits of vermiculture

- The earthworms play a vital role in the entire process; in ploughing and fertilizing the soil and providing all the needed nutrition to the plants.
- The earthworms have contributed to improve the soil structure, soil fertility, promote soil aggregation, encourage favorable soil reactions and enrich the nutrient status of the soil and in the process promoting the plant growth and improving the quality of the produce.
- Earthworms churn the soil and make it porous.
- They improve the soil by helping it achieve proper air, water and solids in the required ratio for maximum plant growth.
- Earthworms improve the water infiltration rate. Its maze of tunnels increases the soil's ability to absorb water.

- Earthworms bring up minerals and make plant nutrients more available.
- Earthworms also neutralize soil pH. Analysis of earthworm castings or manure shows that the soil in the castings has neutral pH (7) regardless of whether the existing soil is above or below pH (7).
- Earthworms compost plant residues.
- Earthworms stimulate microbial population. Free-living nitrogen fixing bacteria are more numerous around the sides of the earthworm's burrows.

Worms consume three times their weight a week or more. Conventional composting takes weeks to months to convert organic material to compost and is very labor intensive.

Using worms to convert organic farm and organic garden waste not only takes far less time than hot composting the material but the vermicompost is far superior to conventional compost. The worm castings in the vermicompost have nutrients that are 97% utilizable by your plants and the castings have a mucous coating which allows the nutrients to "time release".

Using the rich 100% organic vermicompost, which you recycle on site, on your organic farms or organic gardens gives your crops the best fertilizer on the planet.

Problems

- *Odours* : When this occurs it is usually due to the overabundance of "greens" in the bin, which is actually too much nitrogen combining with hydrogen and forms the ammonia. To neutralize the odors you want to add a fair amount of carbon to the mix. The carbon will instead absorb the nitrogen and form a compound that is not smelly. Paper and dried leaves are good sources of carbon. Take note; too much carbon added slows the decomposition process considerably.

- *Pests :* certain types of material, as well as odours from these, can attract pests such as rodents and flies. This is especially true if the loading contains lots of kitchen waste, especially meat. This problem is largely negated if a sealed bin is used where the pests cannot access the material. Most domestic vermicomposters are advised by local authorities to avoid the problem of pests by avoiding using materials that attract them rather than relying on special containers. Ants can become a problem as well. No-see-um netting can be used. Regular mosquito window screen is too large and lets fruitflies and possibly ants in as well.

Precautions

- Vermicompost pit should be protected from direct sun light.
- To maintain moisture level, spray water on the pit as an when required.
- Protect the worms from ant, rat and bird

Dosage

Vermicompost can be applied at 3 to 5 tonnes / ha and for potted plants at 250 g per pot. Dosage can be reduced after one year.

Green Manuring

The practice of turning into the soil undecomposed green plant tissue is referred to as green manuring. Green manure crops can contribute 30 – 60 kg N per ha annually (Greenland, 1986) to the subsequent crop. The cumulative effects of continued use of green manures are important, not only in terms of nitrogen supply but also with regard to soil organic matter and other elements such as phosphate and micro-elements which are mobilised, concentrated in the topsoil and made available for plant growth.

Types of green manuring

Based on the place of cultivation

- Green manuring *in situ* : The green manuring crops are grown in the same field where they are to be incorporated.
- *Green leaf manuring:* If refers to incorporation into soil green leaves and tender twigs collected from shrubs and trees grown outside the field bunds, waste lands, forests, etc.

Based on the form of cultivation

- *Improved fallow:* Replacing natural fallow vegetation with green manure crops to speed up regeneration of soil fertility and permit permanent cultivation, these green manures may be left to grow for one or several years, or only during the dry season.
- *Alley copping:* It is a form of simultaneous fallow in which quickly growing trees, shrubs (usually legumes) or grasses are planted in rows and are regularly cut back; the prunings are used as mulch or worked into the soil in the alleys between the rows.
- *Integration of trees into crop land* : Tree legumes growing among the crops are regularly cut for mulch material to maintain soil fertility in plots.
- *Relay fallow:* Sowing bush legumes among the food crops after these have established and, in the dry season, using the cut green biomass as mulch or working it into the soil; e.g. *Sesbania rostrata*.

- *Live mulching:* In which the rows of food crops are sown into a low but dense cover crop of grasses or legumes, eg. *Centrosema pubescens, Pueraria phaseoloides, Arachis prostrata;* strips of the cover crop are removed by hand or killed by herbicides when the food crops are to be sown, thus reducing soil tillage operations to zero.
- *Shaded green manure:* Practiced in fruit orchards, coffee plots, multistorey kitchen gardens, etc.

Benefits

- Green manuring adds organic matter.
- Green manuring conserves plant nutrients.
- Green manuring increases the availability of plant nutrients.
- Green manuring concentrates plant nutrients in the surface layer of the soil.
- Green manuring improves the structure of the sub soil.
- Green manuring protects the surface soil from beating action of the rain drops and obstructs the flow of water over the surface of the soil.
- Green manuring increases the activity of micro-organisms.
- Green manuring checks weed growth.

Limitations

It cannot be practiced under all conditions and for all crops. There must be an annual rainfall of at least 750 mm or good irrigation facilities to produce sufficient green material and to hasten its decomposition.

Desirable characteristics

- It should be easily established.
- Quick growing especially in the beginning stage.
- It should fix atmospheric nitrogen
- It should have good affinity with mycorrhiza
- It should be efficient water user.
- It should be non-host for crop related pests and diseases.
- It should produce abundant seeds
- It should yield a large quantity of green material.
- It should have more leafy growth than woody growth.
- It should have a deep fibrous root system.
- It should have the ability to grow on poor soils.

Examples

Green manure crops: Sunnhemp, Dhaincha, Beans, Cluster bean, Cowpea, Greengram, Blackgram, Peas, Bersum clover, Soybean, Alfalfa

Green leaf manure crops: Glyricidia, Pongamia, Sesbania, Leucaena.

Integrated Crop Management

To provide an adequate supply of food and other products in an efficient manner. To minimise consumption of non-renewable and other resources. To safeguard the quality of soil, water and the air and to preserve, where feasible, biodiversity in the landscape.To achieve these objectives ICM approach is regarded as the best. According to British Agrochemical Association;

ICM is a method of farming that balances the requirements of running a profitable business with responsibility and sensitivity to the environment. It includes practices that avoid waste, enhance energy efficiency and minimize pollution. ICM combines the best of modern technology with some basic principles of good farming practice and is a whole farm, long term strategy.

ICM is a 'whole farm approach' which is site specific and includes

- The use of crop rotations
- Appropriate cultivation techniques
- Careful choice of seed varieties

Minimum reliance on artificial inputs such as fertilisers, pesticides and fossil fuels Maintenance of the landscape the enhancement of wildlife habitats.

One of the main objectives of ICM is the reduction or replacement of external farm inputs, such as inorganic fertilizers, pesticides and fuel, by means of farm produced substitutes and better management of inputs. Total replacement is not possible without significant loss of yields, but partial substitution of inputs can be achieved by the use of natural resources, the avoidance of waste and efficient management of external inputs. This would then lead to reduced production cost and less environmental degradation. The principles and practices of ICM are outlined below:

Crop rotations

- Increase diversity of crop species to prevent disease and pest carry over from crop to crop
- Ensure effective nutrient uptake by scheduling crops with different nitrogen demands in the correct sequences
- Preserve soil fertility, structure and minimize erosion by ensuring adequate crop cover, good rooting depth and reduction of compaction

Use of disease resistant cultivars to minimize the need for agro chemical inputs

Soil protection

Minimal cultivations to

- Reduce energy usage (i.e. fuel)
- Reduce soil erosion
- Reduce adverse effects on soil invertebrates such as earthworms and predatory beetles and spiders
- However, there should be effective seedbed preparation and crop establishment

Cultivations dependent on soil type, climate and topography of individual farms

Crop nutrition

Nutrient inputs should be carefully balanced in respect of

- Individual crop requirements
- Cropping systems
- Soil residues and residues from previous crop
- Regular soil analysis is recommended
- Use of cover crops/green manures to minimize leaching and erosion

Crop protection

Integrated pest management

- Minimal use of well selected pesticides, i.e. ones that have minimal off target effects
- Alternative husbandry techniques such as mechanical weeding
- In crop monitoring systems (such as traps) to assess pest levels to scale pesticide use to the level of the problem

Improve habitat for predators to increase natural level of biological control

Wildlife and landscape

Planning a programme for the whole farm (cropped and non cropped areas) to enhance biodiversity and landscape features:

Hedges, ditches, field margins, beetle banks and conservation headlands allowing wild species to establish and migrate, and to provide recreational areas for people

A greater diversity of broad leaved weeds may be left within crops to provide food sources for birds and insects, providing the aggressive crop damaging weeds are contained.

Energy efficiency

- Detailed analysis of energy use, especially fossil fuels
- Remedial action to minimize waste
- Consider alternative energy sources
- Change in cultivation practice, i.e. less passes
- Replacement of high fuel consumption machinery, with more efficient alternatives

ICM in practice

The environmental benefits of ICM are difficult to quantify and are related to longer term processes. On the long term projects, biodiversity has increased, there have been improved bird numbers and reduced nitrate leaching and soil erosion. Data from the experimentation, trial farms and various projects has indicated:

Generally a 5 - 15% yield reduction but indications that this is reducing as experience grows.

- Quality of produce is generally maintained
- Variable costs reduced by 20 - 30%
- Gross margins maintained or slightly increased
- Pesticide inputs reduced by 30 - 70%
- Nitrogen inputs reduced by 16 - 25%

The drawbacks

- The control of some weeds is very problematic
- Omission of ploughing and herbicide treatments can lead to a build up of cleavers, blackgrass and other weeds.

❑❑❑

9 TERMINOLOGIES

Agroecology : Agroecology can be defined broadly or narrowly. "Loosely defined, agroecology often incorporates ideas about a more environmentally and socially sensitive approach to agriculture, one that focuses not only on production, but also on the ecological sustainability of the productive system.

or

The science of applying ecological concepts and principles to the design and management of sustainable agroecosystems.

Alternative Farming/Alternative Agriculture : These are essentially synonymous terms encompassing a vast array of practices and enterprises, all of which are considered different from prevailing or conventional agricultural activities. "They include:

- non-traditional crops, livestock, and other farm products;
- service, recreation, tourism, food processing, forest/woodlot, and other enterprises based on farm and natural resources (ancillary enterprises);

- unconventional production systems such as organic farming or aquaculture; or
- direct marketing and other entrepreneurial marketing strategies."

Abiotic factor A non-living component of the environment, such as soil, nutrients, light, fire, or moisture.

Adaptation (i) Any aspect of an organism or its parts that is of value in allowing the organism to withstand the conditions of the environment. (ii) The evolutionary process by which a species' genome and phenotypic characteristics change over time in response to changes in the environment.

Agroecosystem: An agricultural system understood as an ecosystem.

Agroforestry: The practice of including trees in crop-or animal-production agroecosystems.

Allelopathy: An interference interaction in which a plant releases into the environment a compound that inhibits or stimulates the growth or development of other plants.

Alluvium: Soil that has been transported to its present location by water flow (alluvial soil).

Alpha diversity: The variety of species in a particular location in one community or agroecosystem.

Amensalism: An inter-organism interaction in which one organism negatively impacts another organism without receiving any direct benefit itself.

Aquaculture: the production of food and feed using aquatic agroecosystems.

Autotroph: An organism that satisfies its need for organic food molecules by using the energy of the sun, or of the oxidation of inorganic substances, to convert inorganic molecules into organic molecules. Green plants are autotrophs.

Best Management Practices (BMPs) : "... BMPs were developed and implemented as a requirement of the 1977 amendments to the Clean Water Act. BMPs are established soil conservation practices that also provide water quality benefits. They include such practices as cover crops, green manure crops, and stripcropping to control erosion; and soil testing and targeting and timing of chemical applications (similar to IPM) to prevent the loss of nutrients and pesticides. District soil conservation agents use BMPs in helping individual farmers develop conservation plans for their farms".

Biodiversity : "At its simplest level, biodiversity is the sum total of all the plants, animals, fungi and microorganisms in the world, or in a particular area; all of their individual variation; and all the interactions between them".

Agrobiodiversity "is a fundamental feature of farming systems around the world. It encompasses many types of biological resources tied to agriculture, including:

- genetic resources - the essential living materials of plants and animals;
- edible plants and crops, including traditional varieties, cultivars, hybrids, and other genetic material developed by breeders; and
- livestock (small and large, lineal breeds or thoroughbreds) and freshwater fish;
- soil organisms vital to soil fertility, structure, quality, and soil health;
- naturally occurring insects, bacteria, and fungi that control insect pests and diseases of domesticated plants and animals;

- agroecosystem components and types (polycultural/monocultural, small/large scale, rainfed/irrigated, etc.) indispensable for nutrient cycling, stability, and productivity; and
- 'wild' resources (species and elements) of natural habitats and landscapes that can provide services (for example, pest control and ecosystem stability) to agriculture.

"*Agrobiodiversity* therefore includes not only a wide variety of species, but also the many ways in which farmers can exploit biological diversity to produce and manage crops, land, water, insects, and biota".

Biodynamic Agriculture/Biodynamic Farming : Both a concept and a practice, biodynamics "owes its origin to the spiritual insights and perceptions of Dr. Rudolf Steiner, an Austrian philosopher and scientist who lived at the turn of the century". Dr. Steiner emphasized many of the forces within living nature, identifying many of these factors and describing specific practices and preparations that enable the farmer or gardener to work in concert with these parameters. "Central to the biodynamic method... are certain herbal preparations that guide the decomposition processes in manures and compost".

Biointensive Gardening/Mini-farming : John Jeavons and Ecology Action have refined a production system that makes it possible for one person to grow all of his or her family's food using truly sustainable methods that maintain the fertility of the soil without relying on nonrenewable resources like petrochemicals or imported organic matter. The concepts and practices of biointensive gardening were synthesized and introduced to the U.S. by the English master horticulturalist, Alan Chadwick. Important components include double-dug, raised beds; intensive planting; composting; companion planting; and whole system synergy.

Biological Farming/Ecological Farming : Biological and Ecological Farming are terms commonly used in Europe and developing countries. Although sometimes strictly defined, e.g., "Biological farming is a system of crop production in which the producer tries to minimize the use of 'chemicals' for control of crop pests", both biological farming and ecological farming are terms used in the broader sense, encompassing various and more specific practices and techniques of farming sustainability, e.g., organic, biodynamic, holistic, natural.

Norman et al. point to some differentiation between the two terms: "In Europe (e.g., the Netherlands), the term biological often refers to organic farming, whereas the term ecological refers to organic plus environmental considerations such as on-farm wildlife management (i.e., the relationships between parts of the agroecosystem".

Biotechnology : Although farmers have been practicing biotechnology in the broadest sense (i.e. plant and animal breeding to achieve certain traits) for thousands of years, it is the recent breaking of the genetic code that has pushed this science into a new era altogether. Genetic engineering differs significantly from traditional biotechnological techniques in that DNA from different species can be combined to create completely new organisms (Genetically Modified Organisms - GMOs).

Beneficial insects: Beneficial insects are predators, parasites, or competitors of insect pests, helping to regulate pest populations without harm to crops.

β-diversity: The difference in the assemblage of species from one location or habitat to another nearby location or

habitat, or from one part of an agroecosystem to another.

Biogeochemical cycle: The manner in which the atoms of an element critical to life (such as carbon, nitrogen, or phosphorus) move from the bodies of living organisms to the physical environment and back again.

Biogeochemistry: The study of biogeochemical cycling

Biological nitrogen fixation: The fixation, by bacterial cells, of atmospheric Nitrogen gas into organic compounds useful for life. Nitrogen-fixing bacteria exist in the soil and in association and symbiosis with plants or fungi.

Biomass: The mass of all the organic matter in a given system at a given point in time.

Biotic factor: An aspect of the environment related to organisms or their interactions.

Boundary layer: A layer of air saturated with water vapor (from transpiration) that forms next to a leaf surface when there is no air movement.

Buffer zone: A less-intensively-managed and less-disturbed area at the margins of an agroecosystem that protects the adjacent natural system from the potential negative impacts of agricultural activities and management.

Bulk density: The mass of soil per unit of volume.

Carrying Capacity : Carrying capacity is the theoretical equilibrium population size at which a particular population in a particular environment will stabilize when its supply of resources remains constant. It can also be considered to be the maximum sustainable population size; the maximum size that can be supported indefinitely into the future without degrading the environment for future generations.

"The Earth's capacity to support people is determined both by natural constraints and by human choices concerning economics, environment, culture (including values and politics) and demography. Human carrying capacity is therefore dynamic and uncertain. Human choice is not captured by ecological notions of carrying capacity that are appropriate for nonhuman populations.

Conservation Buffer Strips : Conservation Buffer Strips are areas or strips of land maintained in permanent vegetation, designed to intercept pollutants and erosion. Placed around fields, they can enhance wildlife habitat, improve water quality, and enrich aesthetics on farmlands. Various types of buffers include Contour Buffer Strips, Filter Strips, Riparian Forest Buffers, Field Borders, Windbreaks/Shelterbelts, Hedgerows, Grassed Waterways, and Alley Cropping.

Conservation Tillage : Conservation Tillage is a term that covers a broad range of soil tillage systems that leave residue cover on the soil surface, substantially reducing the effects of soil erosion from wind and water. These practices minimize nutrient loss, decreased water storage capacity, crop damage, and decreased farmability. The soil is left undisturbed from harvest to planting except for nutrient amendment. Weed control is accomplished primarily with herbicides, limited cultivation, and, in more sustainable systems, with cover crops.

The National Crop Residue Management Survey (Conservation Technology Information Center (CTIC)) specifies that 30 percent or more of crop residue must be left after planting to qualify as a conservation tillage system. Some specific types of conservation tillage are Minimum Tillage, Zone Tillage, No-till, Ridge-till, Mulch-till, Reduced-till, Strip-till, Rotational Tillage and Crop Residue Management.

Capillary water The water that fills the micropores of the soil and is held to soil particles with a force between 0.3 and 31 bars of suction. Much of this water (that portion held to particles with less than 15 bars of suction) is readily available to plant roots.

Carbonaceous: Rich in carbon.

Carbon dioxide compensation point: The concentration of carbon dioxide in a plant's chloroplasts below which the amount of photosynthate produced fails to compensate for the amount of amount of photosynthate used in respiration.

Carbon fixation: The part of the photosynthetic process in which carbon atoms are extracted from atmospheric carbon dioxide and used to make simple organic compounds that eventually become glucose.

Carbon partitioning: The manner in which a plant allocates to different plant parts the photosynthate it produces.

Catabatic warming: The process that occurs when a large air mass expands after having been forced over a mountain range and becomes warmer and dryer as a result of the expansion.

Cation exchange capacity (CEC): A measurement of a soil's ability to bind positively charged ions (cations), which include many important nutrients. Depends on the amount and type of clay and the amount and *humification* of organic matter in soil. Most of the major cation nutrients are held in soil by CEC (calcium, magnesium, potassium).

Chelation: A reaction between a metallic ion and an organic compound that removes the metallic ion from solution. Chelation is a natural reaction in most soils and is enhanced by organic matter, especially humus. Chelation is similar to cation exchange, except that it usually is more stable in holding ions. Most of the trace cation nutrients are held in soil by chelation (copper, iron, manganese).

Chromosome: Genetic material, composed largely of DNA and protein in the nucleus of cells.

Climax: In classical ecological theory, the end point of the successional process; today, we refer instead to the stage of maturity reached when successional development shifts to dynamic change around an equilibrium point.

C:N ratio: The ratio of carbon to nitrogen in a material. The decomposition of materials is regulated in part by this ratio, so materials with different *C:N ratios* are usually mixed to improve decomposition rates in composting. The *C:N ratio* for optimal biological activity is about 25:1, with higher values being nitrogen limited and lower values being carbon limited. The average C:N ratio for soils is about 10:1.

Colluvium: Soil that has been transported to its present location by the actions of gravity.

Commensalism: An inter-organism interaction in which one organism is aided by the interaction and the other is neither benefited nor harmed.

Community: All the organisms living together in a particular location.

Community Supported Agriculture (CSA): A nationwide movement linking local consumers and farmers into communities. Typical CSA's consist of a group of consumer shareholders that pay a sum in advance in exchange for a regular selection (weekly, biweekly) of in-season crops produced by a farm.

Compensating factor: A factor of the environment that overcomes, eliminates, or modifies the impact of another factor.

Competition: An interaction in which two organisms remove from the environment a limited resource that both require, and both organisms are harmed in the process. Competition can occur between members of the same species and between members of different species.

Composite: A plant in the Compositae (Asteraceae) family.

Compost: Mixed decayed and decaying organic matter with available nutrients useful for fertilizer.

Composting: The management of organic materials to produce *compost*.

Consumer: An organism that ingests other organisms (or their parts or products) to obtain its food energy.

Continental influence: The climatic effect of being distant from the moderating effects of a large body of water.

Coriolis effect: The deflection of air currents in atmospheric circulation cells due to the rotation of the earth.

Cross-pollination: The fertilization of a flower by pollen from the flower of another individual of the same species.

Cultural energy inputs: Forms of energy used in agricultural production that come from sources controlled or provided by humans.

Decomposer: A fungal or bacterial organism that obtains its nutrients and food energy by breaking down dead organic and fecal matter and absorbing some of its nutrient content.

Decomposition: The process by which materials are broken down into simpler compounds by decomposers.

Density-dependent: Directly linked to population density. This term is usually used to describe growth-limiting feedback mechanisms in a population of organisms.

Density-independent: Not directly linked to population density. This term is usually used to describe growth-limiting feedback mechanisms in a population of organisms.

Detritivore: An organism that feeds on dead organic and fecal matter.

Dew point: The temperature at which relative humidity reaches 100% and water vapor is able to condense into water droplets. The dew point varies depending on the absolute water vapor content of the air.

Directed selection: The process of controlling genetic change in domesticated plants through manipulation of the plants' environment and their breeding process.

Disturbance: An event or short-term process that alters a community or ecosystem by changing the relative population levels of at least some of the component species.

Diversity (1) The number or variety of species in a location, community, ecosystem, or agroecosystem. (2) The degree of heterogeneity of the biotic components of an ecosystem or agroecosystem

DNA (Deoxyribonucleic acid): The primary genetic material of life, containing two nucleotide chains in a double-helix.

Domestication: The process of altering, through directed selection, the genetic makeup of a species so as to increase the species' usefulness to humans.

Dominant species: The species with the greatest impact on both the biotic and abiotic components of its community.

Dynamic equilibrium: A condition characterized by an overall balance in the processes of change in an ecosystem, made possible by the system's resiliency, and resulting in relative stability of structure and function despite constant change and small-scale disturbance.

Environmental Indicators: There are diverse interpretations as to what constitute environmental indicators and how they should be used. In any system, however, the "goal of environmental indicators is to communicate information about the environment—and about human activities that affect it—in ways that highlight emerging problems and draw attention to the effectiveness of current policies... an indicator must reflect changes over a period of time keyed to the problem, it must be reliable and reproducible, and, whenever possible, it should be calibrated to the same terms as the policy goals or targets linked to it."

An agri-environmental indicator measures change either in the state of environmental resources used or affected by agriculture, or in farming activities that affect the state of these resources. Examples of sustainable agriculture processes monitored by such indicators are soil quality, water quality, agroecosystem biodiversity, climatic change, farm resource management, and production efficiency.

Easily available water: That portion of water held in the soil that can be readily absorbed by plant roots usually capillary water between 0.3 and 15 bars of suction.

Ecological diversity: The degree of heterogeneity of an ecosystem's or agroecosystem's species makeup, genetic potential, vertical spatial structure, horizontal spatial structure, trophic structure, ecological functioning, and change over time.

Ecological energy inputs: Forms of energy used in agricultural production that come directly from the sun.

Ecological niche: An organism's place and function in the environment, defined by its utilization of resources.

Ecosystem: A functional system of complementary relations between living organisms and their environment within a certain physical area.

Ecotone: A zone of gradual transition between two distinct ecosystems, communities, or habitats.

Ecotype: A population of a species that differs genetically from other populations of the same species because local conditions have selected for certain unique physiological or morphological characteristics.

Edaphic: of the soil, or influenced by the soil. For example, some edaphic factors that influence soil organisms are pH, organic matter content, and hydrology.

Edge effect: The phenomenon of an edge community, or ecotone, having greater ecological diversity than the neighboring communities.

Emergent property: A characteristic of a system that derives from the interaction of its parts and is not observable or inherent in the parts considered separately.

Environmental complex: The composite of all the individual factors of the environment acting and interacting in concert.

Environmental resistance: The genetically-based ability of an organism to withstand stresses, threats, or limiting factors in the environment.

Eolian soil: Soil that has been transported to its current location by the actions of wind (*aeolian* is an acceptable alternative spelling).

Epiphyll: A plant that uses the leaf of another plant for support, but that draws no nutrients from the host plant.

Epiphyte: A plant that uses the trunk or stem of another plant for support, but that draws no nutrients from the host plant.

Evapotranspiration: All forms of evaporation of liquid water from the earth's surface, including the evaporation of bodies of water and soil moisture and the evaporation from leaf surfaces that occurs as part of transpiration.

Field capacity: The amount of water the soil can hold once gravitational water has drained away; this water is mostly capillary water held to soil particles with at least 0.3 bars of suction.

Food and Agriculture Organization: The United Nations organization dedicated to agriculture.

Food demand: The amount of food needed to support a given population or individual for a given amount of time.

Foodsystem: The interconnected meta-system of agroecosystems, their economic, social, cultural, and technological support systems, and systems of food distribution and consumption.

Food security: Can be defined as the "state in which all persons obtain a nutritionally adequate, culturally acceptable diet at all times through local non-emergency sources."

Forb: A non-grass, herbaceous species (generally legumes and composites)

Gene: The basic unit of inheritance comprising a specific sequence of nucleotides on a DNA chain that has a specific function and occupies a specific locus on a chromosome.

Gene flow: The exchange of genetic factors within and between populations by interbreeding or migration.

Generalist: A species that tolerates a broad range of environmental conditions; a generalist has a broad ecological niche.

Genetic engineering: Transfer, by biotechnological methods, of genetic material from one organism to another.

Genetic erosion: The loss of genetic diversity in domesticated organisms that has resulted from human reliance on a few genetically uniform varieties of food crop plants and animals.

Genetic vulnerability: The susceptibility of genetically uniform crops to damage or destruction caused by outbreaks of a disease or pest or unusually poor weather conditions or climatic change.

Genotype: An organism's genetic information, considered as a whole.

Geographic information system (GIS): A computer-based system for managing and analyzing spatially-referenced data. Examples include land-use, topographic and soils data.

Glacial soil: Soil that has been transported to its current location by the movement of glaciers.

Gravitational water: That portion of water in the soil not held strongly enough by adhesion to soil particles to resist the downward pull of gravity.

Green manure: Organic matter added to the soil when a cover crop (often leguminous) is tilled in.

Gross primary productivity: The rate of conversion of solar energy into biomass in an ecosystem.

Geographic information system (GIS): A computer-based system for managing and analyzing spatially-referenced data. Examples include land-use, topographic and soils data.

Glacial soil: Soil that has been transported to its current location by the movement of glaciers.

Gravitational water: That portion of water in the soil not held strongly enough by adhesion to soil particles to resist the downward pull of gravity.

Green manure: Organic matter added to the soil when a cover crop (often leguminous) is tilled in.

Gross primary productivity: The rate of conversion of solar energy into biomass in an ecosystem.

Holistic Management (HM) : "*H*olistic Management is a decision-making process that enables people to make decisions that satisfy immediate needs without jeopardizing their future well-being, or the well-being of future generations. This decision-making process helps people identify their most deeply held values which helps to create clarity in vision and commitment in action. Using that vision to help them create a long-term picture toward which they will progress, people can then use a simple testing process to ensure that the decisions they make will be economically, environmentally, and socially sustainable."

Habitat: The particular environment, characterized by a specific set of environmental conditions, in which a given species occurs.

Hardening: Subjecting a seedling or plant to cooler temperatures in order to increase its resistance to more extreme cold.

Herbaceous: Non-woody.

Herbivore: An animal that feeds exclusively or mainly on plants. Herbivores convert plant biomass into animal biomass.

Heterosis: The production of an exceptionally vigorous and/ or productive hybrid progeny from a directed cross between two pure-breeding plant lines.

Heterotroph: An organism that consumes other organisms to meet its energy needs.

Horizons: Visually distinguishable layers in the soil profile.

Host: An organism that provides food or shelter for another organism.

Host-specificity: The extent to which a parasite is restricted in the host species it can use.

Humification: The decomposition or metabolization of organic material in the soil into humus.

Humus: The fraction of organic matter in the soil resulting from decomposition and mineralization of organic material.

Hybrid vigor: The production of an exceptionally vigorous and/or productive hybrid progeny from a directed cross between two pure-breeding plant lines.

Hydration: The addition of water molecules to a mineral's chemical structure.

Hydrological cycle: The process encompassing the evaporation of water from the earth's surface, its condensation in the atmosphere, and its return to the surface through precipitation.

Hydrology: The properties, distribution, and circulation of water.

Hydrolysis: Replacement of cations in the structure of a silicate mineral with hydrogen ions, resulting in the decomposition of the mineral.

Hydrophyte: A plant that is adapted to grow in water or very wet environments.

Hydroxide clay: A mineral component of the soil without definite crystalline structure composed of hydrated iron and aluminum oxides.

Hygroscopic water: The moisture that is held the most tightly to soil particles, usually with more than 31 bars of suction; it can remain in soil after oven drying.

Integrated Farming Systems (IFS)/Integrated Food and Farming Systems (IFFS) : Farming research and policy programs have begun to recognize that by viewing farms and the food production system as an integrated whole, more efficient use can be made of natural, economic, and social resources. Included in this concept are the goals of finding and adopting "integrated and resource-efficient crop and livestock systems that maintain productivity, that are profitable, and that protect the environment and the personal health of farmers and their families," as well as "overcoming the barriers to adoption of more sustainable agricultural systems so these systems can serve as a foundation upon which rural American communities will be revitalized".

Integrated Pest Management (IPM) : IPM is an ecologically based approach to pest (animal and weed) control that utilizes a multi-disciplinary knowledge of crop/ pest relationships, establishment of acceptable economic thresholds for pest populations and constant field monitoring for potential problems. Management may include such practices as "the use of resistant varieties; crop rotation; cultural practices; optimal use of biological control organisms; certified seed; protective seed treatments; disease-free transplants or rootstock; timeliness of crop cultivation; improved timing of pesticide applications; and removal or 'plow down' of infested plant material".

The term Biointensive IPM emphasizes a "range of preventive tactics and biological controls to keep pest population within acceptable limits. Reduced risk pesticides are used if other tactics have not been adequately effective, as a last resort and with care to minimalize risks".

Biological Control/Bio-control: "Biological control is, generally, man's use of a specially chosen living organism to control a particular pest. This chosen organism might be a predator, parasite, or disease which will attack the harmful insect. It is a form of manipulating nature to increase a desired effect. A complete Biological Control program may range from choosing a pesticide which will be least harmful to beneficial insects, to raising and releasing one insect to have it attack another, almost like a 'living insecticide.'

Integrated nutrient management : It is a strategy to produce and causably increase soil fertility for sustaining crop productivity through optimum use of organic and inorganic sources.

Integrated pest management: It is a strategy in context with farm environment and population dynamics of pests, uses all suitable measures biological, genetic chemical in the most compatible manner so as to maintain pest population below that of causing economic injury.

Intensive/Controlled Grazing Systems : "The term 'intensive grazing' is meant to describe livestock and grass management practices that focus on increased levels of manager involvement, leading to increased productivity and the sustainability of the land. Managers practicing intensive grazing closely follow the interactions between plant, animal, soil and water. They determine where, when and what livestock graze and control animal distribution and movement."

"Controlled grazing is a flexible management method that balances plant and animal requirements. Controlled grazing relies on management, not technology. It uses variable rest periods, short graze periods, high stock densities, and a minimal number of relatively large herds. It requires changing the stocking rate to match annual and seasonal changes in carrying capacity."

Other terms, related to both dairy and meat production, that fall under the category of Intensive/Controlled Grazing are: Rotational Grazing, Management Intensive Grazing (MIG), High-Intensity Low-Frequency Grazing (HILF), Time-Controlled Grazing (TCG), Holistic Range Management, Grassfarming, Pasture-Based Farming, and Voisin Management Grazing.

Importance value: A measure of a species' presence in an ecosystem or community—such as number of individuals, biomass, or productivity—that can be used to determine the species' contribution to the diversity of the system.

Inoculum: The initial organism or organisms that establish a new colony or population; breeding stock. The term is commonly applied to the application of small amounts of microrganisms to establish new populations in soils or composts, for example, Rhizobium bacteria are commonly applied as an inoculum to legume seeds at planting time.

Insolation: The conversion, at the earth's surface, of short wave solar energy into long wave heat energy.

Intermediate disturbance hypothesis: The theory that diversity and productivity in natural ecosystems are highest when moderate disturbance occurs periodically but not too frequently.

Intercropping: Planting more than one crop in a field using a regular pattern that interleaves each crop in some pattern. A form of polyculture.

Interspecific competition: Competition for resources among individuals of different species.

Intraspecific competition: Competition for resources among individuals of the same species.

Inversion: The sandwiching of a layer of warm air between two layers of cold air in a valley.

Integrated Pest Management: Pest control using an array of complementary approaches including natural predators, parasites, pest-resistant varieties, pesticides, and other biological and environmental control practices.

K-strategist A species that lives in conditions where mortality is density-dependent; a typical *K*-strategist has a relatively long lifespan and invests a relatively large amount of energy in each of the few offspring it produces.

Local/Community Food System : A community food system, also known as a local food system, "is a collaborative effort to integrate agricultural production with food distribution to enhance the economic, environmental, and social well-being of a particular place (i.e. a neighborhood, city, county or region)".

"One of the primary assumptions underlying the sustainable diet concept is that foods are produced, processed, and distributed as locally as possible. This approach supports a food system that preserves local farmland and fosters community economic viability, requires less energy for transportation, and offers consumers the freshest foods".

Low Input Agriculture : Low input farming systems "seek to optimize the management and use of internal production inputs (i.e. on-farm resources)... and to minimize the use of production inputs (i.e. off-farm resources), such as purchased fertilizers and pesticides, wherever and whenever feasible and practicable, to lower production costs, to avoid pollution of surface and groundwater, to reduce pesticide residues in food, to reduce a farmer's overall risk, and to increase both short- and long-term farm profitability."

The term is "somewhat misleading and indeed unfortunate. For some it implied that farmers should starve their crops, let the weeds choke them out, and let insects clean up what was left. In fact, the term low-input referred to purchasing few off-farm inputs (usually fertilizers and pesticides), while increasing on-farm inputs (i.e. manures, cover crops, and especially management). Thus, a more accurate term would be different input or low external input rather than low-input."

Landrace: A locally-adapted strain of a species bred through traditional methods of directed selection. Also called *farmer varieties.*

Landscape ecology: The study of environmental factors and interactions at a scale that encompasses more than one ecosystem at a time.

Land Utilization Type (LUT): A kind of land use described or defined in a degree of detail greater than that of a major kind of land use. In the context of irrigated agriculture, a land utilization type refers to a crop, crop combination or cropping system with specified irrigation and management methods in a defined technical and socio-economic setting. In the context of rainfed agriculture, a land utilization type refers to a crop, crop combination or cropping system with a specified technical and socio-economic setting. A land utilization type in forestry consists of a technical specifications in a given physical, economic and social setting.

Legume: Any of the leguminaceae family that may tree, shrubs and herbs many of which have capacity to live in symbiotic relationship with rhizobia

Ley farming: It is the alteration of food crops with pastures on same piece of land.

Litter: Uppermost layer of organic matter on soil surface including leaves, twigs, flowers freshly fallen or slightly decomposed.

Leaf-area index: A measure of leaf cover above a certain area of ground, given by the ratio of total leaf surface area to ground surface area.

Legume: A plant in the Leguminosae (Fabaceae) family. Most species in this family can fix nitrogen.

Light compensation point: The level of light intensity needed for a plant to produce an amount of photosynthate equal to the amount it uses for respiration.

Light reactions: The components of photosynthesis in which light energy is converted into chemical energy in the form of ATP and NADPH.

Lignin: an amorphous polymer related to cellulose that cements cell walls, helping them stay rigid. Lignin is highly resistant to decomposition.

Limiting nutrient: A nutrient not present in the soil in sufficient quantity to support optimal plant growth.

Living mulch: A cover crop that is interplanted with the primary crop(s) during the growing season.

Lodging: The flattening of a crop plant or crop stand by strong wind, usually involving uprooting or stem breakage.

LEISA: It is a system of agriculture in which most of the inputs used originate from own farm or a region and action is taken to ensure sustainability.

Macronutrient: A nutrient plants need in large quantities; the macronutrients include carbon, nitrogen, oxygen, phosphorus, sulfur, and water.

Maritime influence: The moderating effect of a nearby large body of water, such as an ocean, on the weather and climate of an area.

Mass selection: The traditional method of directed selection, in which seed is collected from those individuals in a population that show one or more desirable traits and then used for planting the next crop.

Mesophyte: A plant that is adapted to environments that are neither very dry nor very wet. Compare with *xerophyte* and *hydrophyte*.

Microclimate: The environmental conditions in the immediate vicinity of an organism.

Micronutrient: A nutrient necessary for plant survival but needed in relatively small quantities.

Mineralization: The process by which organic residues in the soil are broken down to release mineral nutrients that can be utilized by plants.

Mountain wind: The down slope movement of air at night that occurs as the upper slopes of a mountain cool more rapidly than those below.

Multiple cropping: The cultivation of two or more crops in succession or with some overlap in the same field within one year. Double-cropping of rice after wheat is an example. When crops overlap in time, multiple cropping is a form of polyculture.

Mutualism: An interaction in which two organisms impact each other positively; neither is as successful in the absence of the interaction.

Mycorrhizae: Symbiotic fungal connections with plant roots through which a fungal organism provides water and nutrients to a plant and the plant provides sugars to the fungi.

Natural Farming : Natural Farming reflects the experiences and philosophy of Japanese farmer Masanobu Fukuoka. His books The One-Straw Revolution: An Introduction to Natural Farming describe what he calls "do-nothing farming" and a lifetime of nature study. "His farming method involves no tillage, no fertilizer, no pesticides, no weeding, no pruning, and remarkably little labor! He accomplishes all this (and high yields) by careful timing of his seeding and careful combinations of plants (polyculture). In short, he has brought the practical art of working with nature to a high level of refinement."

Networking: Establishing and strengthening links between individual groups and organizations with similar interest and objective.

Non-renewable resources: Resources such as oil, coal and mineral host which can not be regenerated on time scale relevant to human and are exhaustible

Nutrient cycling: The recurring flow of nutrients through farm system.

Nature Farming: Nature farming grew out of the philosophy and methodology of Japanese philosophist, Mokicho Okada in the mid-1940s. "The theory of Nature Farming, as Okada expounded it, rests on a belief in the universal life-giving powers that the elements of fire, water, and earth confer on the soil... The planet's soil, created over a span of eons, has acquired life-sustaining properties, in accordance with the principle of the indivisibility of the spiritual and the physical realms, which in turn provide the life-force that enables plants to grow. To utilize the inherent power of the soil is the underlying principle of Nature

Farming". Practices focus on analyzing and building soil through composting, green manuring, mulch, and various other soil management techniques. Similar in many ways to organic farming, nature farming is most commonly practiced in the Pacific Rim countries of Asia and North America.

Kyusei Nature Farming: Developed by Teruo Higa in Japan during the 1980s, "Kyusei Nature Farming means saving the world through natural or organic farming methods. An added dimension of Kyusei Nature Farming is that it often employs technology involving beneficial microorganisms as inoculants to increase the microbial diversity of agricultural soils, which, in turn, can enhance the growth, health, and yield of crops."

Natural selection: The process by which adaptive traits increase in frequency in a population due to the differential reproductive success of the individuals that possess the traits.

Net primary productivity (NPP): The difference between the rate of conversion of solar energy into biomass in an ecosystem and the rate at which energy is used to maintain the producers of the system.

Niche amplitude: The size or range of one or more of the dimensions of the multidimensional space encompassed by a particular species' niche. The niche amplitude of a generalist species is larger than that of a specialist species.

Niche breadth: Essentially a synonym for niche amplitude.

Niche diversity: Differences in the resource-use patterns of similar species that allow them to coexist successfully in the same environment.

Non-governmental organization (NGO): An organization that is not part of a government. Private foundations and development organizations are examples.

Nucleotide: A sub unit of DNA or RNA molecules, part of the genetic code.

Nutrient Management: Nutrient management is "managing the amount, source, placement, form, and timing of the application of nutrients and soil amendments to ensure adequate soil fertility for plant production and to minimize the potential for environmental degradation, particularly water quality impairment."

Nutrient management has taken on new connotations in recent times. Soil fertility traditionally dealt with supplying and managing nutrients to meet crop production requirements, focusing on optimization of agronomic production and economic returns to crop production. Contemporary nutrient management deals with these same production concerns, but recognizes that ways of farming must now balance the limits of soil and crop nutrient use with the demands of intensive animal production. Current decision-making processes include crop and animal production factors, economic factors, and the integrity of local surface water and groundwater, as well as the fate of far-away environmental systems.

Organic Farming : The term 'organic farming' was first used by Lord Northbourne in the book, *Look to the Land* (London: Dent, 1940. NAL Call # 30 N81). Lord Northbourne, who embraced the teachings of Rudolph Steiner and biodynamic farming, had a "vision of the farm as a sustainable, ecologically stable, self-contained unit, biologically complete and balanced – a dynamic living organic whole. The term thus did not refer solely to the use of living materials (organic manures, etc) in agriculture although obviously it included them, but with its emphasis on 'wholeness' is encompassed best by the definition 'of, pertaining to, or characterized by systematic connexion or coordination of parts of the one whole.' (Oxford English Dictionary, 1971.)"

As defined by a USDA Study Team on Organic Farming, "Organic farming is a production system which avoids or largely excludes the use of synthetically compounded fertilizers, pesticides, growth regulators, and livestock feed additives. To the maximum extent feasible, organic farming systems rely upon crop rotations, crop residues, animal manures, legumes, green manures, off-farm organic wastes, mechanical cultivation, mineral-bearing rocks, and aspects of biological pest control to maintain soil productivity and tilth, to supply plant nutrients, and to control insects, weeds and other pests."

The following definition was drafted and passed by the USDA National Organic Standards Board (NOSB) in April 1995. It was developed by a joint NOSB/ National Organic Program task force, and incorporated language from the Codex Draft Guidelines for organically produced foods: "Organic agriculture is an ecological production management system that promotes and enhances biodiversity, biological cycles and soil biological activity. It is based on minimal use of off-farm inputs and on management practices that restore, maintain and enhance ecological harmony. 'Organic' is a labeling term that denotes products produced under the authority of the Organic Foods Production Act.

"The principal guidelines for organic production are to use materials and practices that enhance the ecological balance of natural systems and that integrate the parts of the farming system into an ecological whole. Organic agriculture practices cannot ensure that products are completely free of residues; however, methods are used to minimize pollution from air, soil and water. Organic food handlers, processors and retailers adhere to standards that maintain the integrity of organic agricultural products. The primary goal of organic agriculture is to optimize the health and productivity of interdependent communities of soil life, plants, animals and people".

Organic Agriculture: Organic farming lies at the opposite end of a spectrum of approaches from intensive agriculture, but has some radical distinctions from all the others. While it shares some similar practical aims with integrated farming, organic agriculture it is qualitatively different. It is not so much about not using syntheric pesticides and fertilizers, as representing a different philosophy of agriculture. It is a holistic, integrated system of farming based on the management of the natural ecosystem and its processes, so that crop production is in balance with the ability of the soil to release nutrients. Synthetic fertilizers are unnecessary because of the way organic methods manage the soil, its microbes and nutrients, as summed in the phrase feed the soil and let the soil feed the plant. Indeed, soluble inorganic fertilizers which directly feed the plant are seen as impoverishing the soil. Instead of fertilizers, wastes from crop residues and manure become nutrients for crop products. Thus unlike intensive arable systems, most organic farming includes livestock as an integral part of the system. Rotating the use of a field with carefully chosen species enables nutrients to be recycled within the system. Crop health is to be maintained by natural systems of pest, weed and disease control, using rotations, mechanical weeding and various ingenous planting patterns which explit the natural pest and diseases resistance of certain plant combinations.

Open pollination: The natural dispersal of pollen among all the members of a cross-pollinating crop population, resulting in the maximum degree of genetic mixing and diversity.

Organic matter: Material containing molecules based on Carbon, usually referring to soil organic matter.

Organism: An individual of a species.

Overyielding: The production of a yield by an intercrop that is larger than the yield produced by planting the component crops in monoculture on an equivalent area of land.

Oxidation: The loss of electrons from an atom that accompanies the change from a reduced to an oxidized state.

Output: Products or function, which are obtained by means of farming activities and consumed by, farm household or reinvested in farming or may be enhanced or sold.

Participatory technology development: It is process of combining the indigenous knowledge and research capacities of local farming communities with that of research and development institution in an interactive way in order to identify generate test and apply new techniques to strengthen the existing technology.

Permaculture : A contraction of "permanent agriculture," the word "permaculture" was coined by Australian Bill Mollison in the late 1970s. One of the many alternative agriculture systems described as sustainable, permaculture is "unique in its emphasis on design; that is, the location of each element in a landscape, and the evolution of landscape over time. The goal of permaculture is to produce an efficient, low-maintenance integration of plants, animals, people and structure... applied at the scale of a home garden, all the way through to a large farm".

Precision Farming/Agriculture : Precision agriculture is a "management strategy that employs detailed, site-specific information to precisely manage production inputs. This concept is sometimes called Precision Agriculture, Prescription Farming, Site-specific Management. The idea is to know the soil and crop characteristics unique to each part of the field, and to optimize the production inputs within small portions of the field. The philosophy behind precision agriculture is that production inputs (seed, fertilizer, chemicals, etc.) should be applied only as needed and where needed for the most economic production."

This system requires the utilization of sophisticated technology including personal computers, telecommunications, global positioning systems (GPS), geographic information systems (GIS), variable rate controllers, and infield and remote sensing. Although precision agriculture promises reduced use of chemical inputs, there are several factors that make it controversial in the sustainable agriculture community, including the requirements of large capital outlay and advanced technical expertise.

Parasite: An organism that uses another organism for food and thus harms the other organism.

Parasitism: An interaction in which one organism feeds on another organism, harming (but generally not killing) it.

Parasitoid: A parasite that feeds on predators or other parasites.

Patchiness: A measurement of the diversity of successional stages present in a specific area.

Patchy landscape: A landscape with a diversity of successional stages or habitat types.

Percolation: Water movement through the soil due to the pull of gravity.

Permanent wilting point: The level of soil moisture below which a plant wilts and is unable to recover.

pH: The logarithm of the reciprocal of the concentration of hydrogen ions in a medium, like water or soil (log10{1/[H+]}). pH values range from 0 to 14, giving the relative acidity or alkalinity of a medium, with a pH of 7 being neutral, and lower values being acidic, higher values, alkaline.

Phenotype: The physical expression of the genotype; an organism's physical characteristics.

Photoperiod: The total number of hours of daylight.

Photorespiration: The energetically-wasteful substitution of oxygen for carbon dioxide in the dark reactions of photosynthesis, which occurs when plant stomata close and carbon dioxide concentration declines.

Photosynthate: The simple-sugar end products of photosynthesis.

Polyculture: Cropping systems in which different crop species are grown in mixtures in the same field at the same time.

Population: A group of individuals of the same species that live in the same geographic region.

Potential niche: The maximum possible distribution of a species in the environment.

Predation: An interaction in which one organism kills and consumes another.

Predator: An animal that consumes other animals to satisfy its nutritive requirements.

Prescribed burn: A fire set and controlled by humans to achieve some management objective, such as improving pasture in grazing systems.

Prevailing winds: The general wind patterns characteristic of broad latitudinal belts on the earth's surface.

Primary production: The amount of light energy converted into plant biomass in a system.

Primary succession: Ecological succession on a site that was not previously occupied by living organisms.

Producer: An organism that converts solar energy into biomass.

Production: Harvest output or yield.

Productivity index: A measure of the amount of biomass invested in the harvested product in relation to the total amount of standing biomass present in the rest of the system.

Productivity: It can be expressed as output per unit land, capital, labour, energy, water, nutrients etc.

Productivity: The ecological processes and structures in an agroecosystem that enable production.

Protein: A compound formed from a chain of amino acids. Proteins are present in all living things, and are used for enzymes, hormones and other essential molecules.

Protocooperation: An interaction in which both organisms are benefited if the interaction occurs, but neither are harmed if it does not occur. Polyculture is agriculture using multiple crops in the same space, in imitation of the diversity of natural ecosystems, and avoiding large stands of single crops, or monoculture. It includes crop rotation, multi-cropping, inter-cropping, and alley cropping.

Regenerative Agriculture : Robert Rodale coined this term, and it subsequently was expanded to "regenerative/ sustainable agriculture" by the Rodale Institute and Rodale Research Center. Two reasons given for the emphasis on "regenerative" are (1) "enhanced regeneration of renewable resources is essential to the achievement of a sustainable form of agriculture," and (2) "the concept of regeneration would be relevant to many economic sectors and social concerns."

Stability: It is collective aspect of the characteristics of system that will minimize negative effects of abrupt changes.

Sustainable Agriculture Research and Education (SARE) program (USDA) : SARE is the U.S. Department of Agriculture's primary means of studying and publicizing sustainable agriculture practices. Through a competitive grants program that works with teams of agencies, organizations, and farmers, more than 1200 projects have been implemented.

Initially called the Low-Input Sustainable Agriculture (LISA) program, SARE was authorized by Congress in the Food Security Act of 1985 (P.L. 99-198) in response to widespread acknowledgment that science-based information was lacking for farmers seeking to reduce chemical use in crop production. The LISA program got started with a $3.9 million appropriation in 1988. The Food, Agriculture, Conservation and Trade Act of 1990 renamed LISA the Sustainable Agriculture Research and Education Program, and added two other programs—one for research on integrated crop/livestock operations, and another to train Extension Service agents in disseminating sustainable farming practices. In 1991, the SARE program began cooperating with the Environmental Protection Agency to administer the Agriculture in Concert with the Environment (ACE) program.

Sustainable Development: Sustainable development must. "meet the needs of the present without compromising the ability of future generations to meet their own needs."

Subsistence agriculture: Farming system in which large part of final yield is consumed by producer.

Synergy: The action of two or more substances, organisms to achieve an effect of which each is individually incapable.

Whole Farm Planning: Whole farm planning strategies share a conservation, family-oriented approach to farm management, although specific components may vary from farm to farm, and from community to community. "Whole farm planning provides farmers with the management tools they need to manage biologically complex farming systems in a profitable manner. As a management system, it draws on cutting-edge management theory used by other businesses, industries and even cities. It encourages farmers to set explicit goals for their operation; carefully examine and assess all the resources – cultural, financial, and natural – available for meeting their goals; develop short- and long-term plans to meet their goals; make decisions on a daily basis that support their goals; and monitor their progress toward meeting goals".

❑❑❑

SELECTED REFERENCES

Alien, P., D.V. Dusen, J. Lundy, and S.R. Gliessman. (1991), Integrating social environmental, and economic issues in sustainable agriculture. American Journal of Alternative Agriculture 6:34-39.

Altieri, M.A. (1987). Agroecology: The Scientific Basic of Alternative Agriculture 2nd ed. Westview Pr., Boulder, Colorado.

Altieri, M,A., D.K. Letourneaour and J.R. Davis. (1993). Developing sustainable agroecosystems. Bioscience 33: 45-49.

Altieri, M.A, (1995), Biodiversity and pest management in agroecosystems. Haworth Press, New York.

Allied, M.A. (1995). Biodiversity, ecosystem function and insect pest management in agricultural systems. In W.W. Collins and C.O. Qualset (eds.) Biodiversity in Agroecosystems. CRC Press. Boca Raton.

Audsley, R. (ed), Alber, S., Clift, R., Cowcll. S., Gaillard, Hausheer, J., Jolliett, O., Kleijn, R., Mortensen, B., Pearce, D., Roger, E., Teulon, H., Weidema, B. and van Zeits, H, (1997). Harminisation of environmental life cycle assessment for agriculture, Final report.

European commission concerted action AIR3-CT 94-2028. Silsoe Research Institute. Bedford, UK.

Barlow, K. E. (1942). The Discipline of peace. Faber and Faber, London, England. 214 pp.

Bartelmus, Peter (1997). "Whither Economics? From Optimality to Sustainability" Environment and Development Economics, Vol. 2, Part 3.

Bill Mollison, Permaculture: A Designer's Manual, Tagari Press, (1988).

Blaser, Peter, and Ehrenfried Pfeiffer (1984). Bio-Dynamic Composting on the Farm; How Much Compost Should We Use? Biodynamic Farming and Gardening Association, Inc., Kimberton, PA. 23p.

Borlaug, N. E. and C. R Dowswell (1994). Feeding a human population that increasingly crowds a fragile planet. Keynote lecture, 15th World Congress of Soil Science, Acapulco, Mexico. 15 p.

Brown, L. R., *et al.* (1993), State of the world 1993, W. W. Norton, London and New York.

Browne, D. (1855). The field book of manures or the American mulch book. C. M. Saxton and Co., New York, New York. 422 pp.

Carroll, C.R., J.H. Vandermeer and P. Rosset, editors. (1990). Agroecology. McGraw-Hill Publications, New York.

Connell, O., Sustainable Agriculture, A Valid Alternative, 1992, Outlook on Agriculture 21(1): p.6

Dalgaard, T., J.R. Porter, and N. Hutchings (2002). Agroecology, scaling, and interdisciplinarity. European Journal of Agronomy (in review)

David Norman *et al.*, (1997). Defining and Implementing Sustainable Agriculture, Kansas Sustainable

Agriculture Series, Paper #1; Manhattan KS: Kansas Agricultural Experiment Station.

Doherty, S. and Rydberg, T. (ed), Ekblad, G., Gronlund, E., Ingamarsson, F., Karlsson, L., Nolsson, S. and Strid Eriksson, I (2002). Ecosystem Properties and Principles of living Systems as Foundation for Sustainable Agriculture -Critical Reviews of environmental assessment tools, key findings and questions from a course process. Report Ekologiskt lantbruk, No 32. Center for Ecological Agriculture, SLU. Uppsala, Sweden.

Doran, J. W., D. C. Coleman, D. F. Bezdicek, and B. A. Stewart (eds)., (1994). Defining Soil Quality for a Sustainable Environment. SSSA Special Pub. 25. SSSA-ASA. Madison, Wise.

Dover, M.S., and L.M Talbot, (1987). To Feed the Earth : Agro-Ecology for Sustainable Development. World Resources Institute, Washington D.C.

Dregne, H, 1990. Desertification of Drylands, in challenges in Dryland Agriculture – A global perspective. Proc. of the Intt. Conf. on Dryland Fmg., Amarillo/ Bushland, Taxas, USA, 15 - 18 August 1988.

Edwards, R Lal, P Madden. RH Miller, and GJ House. Ankeny IA: (1990). Sustainable Agricultural Systems, Soil and Water Conservation Society.

Elliot, R.H. (1907). The Clifton Park System of Farming. Faber and Faber, London, England. 261 pp.

Fall, Cox, Caroline "Biotechnology and agricultural pesticide use: an interaction between genes and poisons."Journal of Pesticide Reform 13(3):4-11.

Fletcher, S. (1907). Soils: How to handle and improve them. Doubleday, Page and Co., New York, 438 pp.

Fukuoka, M. (1985). The Natural Way of Farming: the theory and practice of green philosophy. Japan Publications, Tokyo, Japan.

Gliessman, S.R., editor (1978). Agroecosistemas con enfasis en el estudio de tecnologia agricola tradicional. Colegio Superior de Agricultural Tropical, Gardenas, Mexico.

Gliessman, S.R., editor (1990), Agroecology: Researching the Ecological Basis for Sustainable Agriculture. Springer-Verlag, New York.

Gliessman, S.R. (1998). Agroecology: ecological processes in sustainable agriculture. Ann Arbor Press, Michigan.

Gliessman, S.R., editor (2001). Agroecosystem Sustainability: Developing Practical Strategies. CRC Press, Boca Raton, Florida.

Hill, Julie, 1992, "An Environmentalist's perspective on genetic modification." Biotechnology Education 3(2):52-54.

Jackson, W. 1980, New roots for Agriculture. Friends of the Earth, University of Nebraska Press, Lincoln, Nebraska.

Johansson, S., Doherty S. & Rydberg, T. 2000, Sweden Food System Analysis. Proceedings to the First Biennial Energy Research Conference: Energy Quality and Transformities, September 2-4, 1999, University of Florida, Gainesville.

Keeney, D. R. 1997a, What goes around comes around: the nitrogen issues cycle, Proceedings. 1997. International Symposium on Fertilizers and the Environment. Haifa, Israel.

Lal, R. and Miller, F. P. 1990, Sustainable Farming Systems for the Tropics, in Sustainable Agriculture - Issue,

Perspectives and Prospects in Semi Arid Tropics, Proc. Intl. Symp. on Natural Resources Management for a Sustainable Agriculture, Vol. 1. p. 69-89, Ind. Soc. Agron.

Lampkin, N. H. (1999). Organic farming in the European Union: overview, policies and perspectives. Invited keynote paper in: Organic Farming in the European Union. Perspectives for the 21st Century. Proceedings of EU Commission/Austrian Government Conference, Baden/Vienna, May (1999). EuroTech Management, Vienna, pp. 23-30.

Lockeretz, William and Fall, 1988, "Open questions in sustainable agriculture." (commentary). American Journal of Alternative Agriculture 3(4):174-181.

Mclsaac G and WR Edwards., (1994). Sustainable Agriculture in the American Midwest: Lessons From the Past, Prospects for the Future, University of Illinois Press.

Moench, Marcus and Dinesh Kumar (1997), "Distinction between efficiency and sustainability: The role of energy prices in groundwater management, in Anil Agarwal (Ed.) (1997), The Challenge of the Balance: Environmental Economics in India, Centre for Science and Environment, New Delhi.

Mollison, B. (1979). Permaculture Two. Tagari, Stanley, Tasmania.

Nadkarni, M.V. (1996), "Forests, People and Economics", Indian Journal of Agriculture Economics, Vol. 51. Nos. 1 and 2, January-June.

Northburn, Lord (1940). Look to the land. Dent, London, England.

Odum, H.T. (1996). Environmental Accounting - Energy and Environmental Decision Making. John Wiley & Sons, Inc. New York.

Oakeshott, J.G.and Whitten, M.J., (1991). "Opportunities for modern biotechnology in control of insect pests and weeds, with special reference to developing countries." FAO Plant Protection Bulletin 39(4): 155-181.

Parr, J. F., Stewart, B. A., Hornick, S. B., and Singh, R. P. (1990). Improving the Sustainability of Dryland Farming System : A Global Perspective. Adv. Soil Sci., 13, 1-8.

Pezzey J, (1992). Sustainable Development Concepts: An Economic Analysis, Washington DC: World Bank.

Pimental, D. MS Hunter, JA LaGro, RA Efroymson, JC Landers, FT Mervis," CA McCarthy, and AE Boyd., (1989). Benefits and Risks of Genetic Engineering in Agriculture, BioScience 39 (9): pp. 606-614.

Pretty, J.N. (1994). Regenerating Agriculture. Earthscan Publications Ltd., London.

Reijntjes, C.B., Haverkort and A. Waters-Bayer., (1992), Farming for the future. MacMillan Press Ltd., London.

Roberts, I (1907). The fertility of the land. Macmillan Publishing Co., New York, 415pp.

Rodal, K. (J983). Breaking in Ground: The search for a SustainabJe Agriculture. The Futurist 1(1): 15-20.

Soule, J.D., and J.K. Piper., (1992). Farming in Nature's Image: an Ecological Approach to Agriculture. Island Press, Washington, DC.

Steiner, Rudolf, (1993). Spiritual Foundations for the Renewal of Agriculture: A Course of Lectures. Bio-Dynamic Farming and Gardening Association., Kimberton, PA. 310 p.

Torjusen, H.G. Lieblein, M. Wandel, and C.A. Francis., (2001). Food system orientation and quality perception among consumers and producers of organic food in Hedmark County, Norway. Food Quality and Preference 12:207-21.

Vandermeer, J. (1995). The Ecological Basis of Alternative Agriculture. Annual Review of Ecological Systems 26:201-224.

Wackernagel, M., Onisto, L., Bello, P., Callejas Linares, A., Lopez Falfan, I. S., Mendez Garcia, J., Suarez Guerrero, A. I. And Snarez Guerrero, M. G. (1999). National Natural Capital Accounting with the Ecological Footprint Concept. Ecol. Econ. 29:375-390.

Waksman, S. (1936). Humus: Origin, chemical composition and importance in nature. Williams and Wilkins Co., Baltimore, Maryland. 526 pp.

❑❑❑

Zeitfracht Medien GmbH
Ferdinand-Jühlke-Straße 7
99095 Erfurt, Deutschland
produktsicherheit@kolibri360.de